Silverlight 应用与开发

罗俊海 郑龙 周忠宝 编著

清华大学出版社

北京

内 容 简 介

本书从 Silverlight 程序设计初学者的角度出发，对 Silverlight 语言的概念和技术等基础内容进行了全面、详细的讲解。全书共含 6 章理论和 6 个上机实训，主要包括 Silverlight 概述，矢量绘图、画刷与着色，图像与视觉特效，动画与多媒体，Silverlight 与 HTML、JavaScript 三者交互，数据访问与 Silverlight 高级应用实例，每章都配有丰富的实例、要点和作业，帮助读者理解和掌握书中的内容。

本书适合作为计算机相关专业"Silverlight 程序设计"课程的培训教材，也可作为程序设计员或对 Silverlight 编程感兴趣的读者的入门参考书，还可供面向对象编程爱好者和自学 Silverlight 编程的读者使用。

本书封面贴有清华大学出版社防伪标签，无标签者不得销售。
版权所有，侵权必究。侵权举报电话：010-62782989 13701121933

图书在版编目(CIP)数据

Silverlight 应用与开发/罗俊海，郑龙，周忠宝编著.—北京：清华大学出版社，2018
ISBN 978-7-302-50409-2

Ⅰ.①S… Ⅱ.①罗…②郑…③周… Ⅲ.①网页制作工具 Ⅳ.①TP393.092.2

中国版本图书馆 CIP 数据核字(2018)第 123038 号

责任编辑：文　怡
封面设计：台禹微
责任校对：梁　毅
责任印制：董　瑾

出版发行：清华大学出版社
网　　址：http://www.tup.com.cn，http://www.wqbook.com
地　　址：北京清华大学学研大厦 A 座
邮　　编：100084
社 总 机：010-62770175
邮　　购：010-62786544
投稿与读者服务：010-62776969，c-service@tup.tsinghua.edu.cn
质量反馈：010-62772015，zhiliang@tup.tsinghua.edu.cn
课件下载：http://www.tup.com.cn，010-62795954

印 装 者：北京嘉实印刷有限公司
经　　销：全国新华书店
开　　本：185mm×260mm　　印　张：13.75　　字　数：334 千字
版　　次：2018 年 10 月第 1 版　　印　次：2018 年 10 月第 2 次印刷
定　　价：49.00 元

产品编号：078230-01

PREFACE

时光荏苒,如白驹过隙,一转眼中国互联网已走过了30年历程。回首过去,人工智能、云计算、移动支付这些互联网产物不仅迅速进入了我们的生活,刷新了我们对科技发展的认知,也提高了我们的生活质量水平。人们谈论的话题已经离不开这些,例如:人工智能是否会替代人类,成为工作的主要劳动力;数字货币是否会代替纸币流通于市场;虚拟现实体验到底会有多真实多刺激;就连开滴滴的司机师傅都会在人机围棋大战的赛事上与你赌一把。从这些现象中不难发现,互联网的辐射面在不断变广,计算机科学与信息技术发展的普适性在不断变强,信息技术如化雨春风,润物无声地全面融入,颠覆了我们的生活。

1987年,我国网络专家钱天白通过拨号方式在国际互联网上发出了中国有史以来第一封电子邮件"越过长城,走向世界",从此,我国互联网时代开启。30年间,人类社会仍然遵循着万物生长规律自然成长,但互联网的枝芽却依托人类的智慧于内部结构中野蛮扩延,并且主流设备、主流技术的迭代速度越来越快。目前,人们的生活状态是"拇指在手机屏幕方寸间游走的距离,已经超过双脚走过的路程"。

截至2017年6月,中国网民规模已达到7.5亿,占全球网民总数的1/5,而且这个数字还在不断地增加。这是一个巨大的互联市场,可以得到我们所需要的内容:有可能是一个简单的Web页面,也有可能是一个复杂的应用程序。

然而,面对快速发展的互联网,每一个互联网人亦感到焦虑,感觉它运转的速度已经快到我们追赶的极限。信息时刻在更新,科技不断被颠覆,想象力也一直被挑战,面对这些,人们感到不安的同时又对未来的互联网充满期待。

互联网的魅力正在于此,恰如山之两面,一面阴暗晦涩,另一面生机勃勃,一旦跨过山之巅,即是不一样的风景。这样的挑战会让人着迷,并甘愿为之付出努力。而这个行业还有很多伟大的事情值得去琢磨,去付出自己的匠心。

本书从Silverlight程序设计初学者的角度出发,对Silverlight语言的概念和技术等基础内容进行了全面、详细的讲解。对书中内容所涉及的知识点和相关信息,应了解、掌握,夯实基础,切不可急于求成;有相关经验、但了解不足的开发人员,也可从本书中找到许多不同领域的兴趣点和用法。本书实例内容选取市场流行应用项目或产品项目,并附有章后练习题,部分练习题模拟大型软件开发企业实例项目,比较具体,其他练习题则较为通俗易懂,旨在提高读者对相应章节内容的理解程度,帮助读者巩固本章的内容。

本书在编写过程中获得了国家自然科学基金委员会与中国民用航空局联合资助项目(U1733110)、中央高校基本科研业务费专项资金(2672018ZYGX2018J018)、湖南省科学"十三五"规划课题(XJK016BGD009)、湖南省教学改革研究课题(2015001)、湖南省自然科学基金(2017JJ1012)、国家自然科学基金(71371067)的资助,并得到了电子科技大学、湖南大学、国防科技大学、佛山科学技术学院和深圳华大乐业教育科技有限公司等领导的大力支持,同时参考了一些相关著作和文献,在此深表感谢。

  本书撰写过程中得到了何敏藩、邢立宁、姚锋、叶昭晖、邓劲生、姚煊道、邹伟、王浩、张章、肖丹、蔡琴、付艳和周滔等编委老师的帮助,在此向这些老师深表感谢。

  未来互联网信息技术已扑面而来,汹涌胜于往昔,你做好准备了吗?

编 者
2018 年 6 月

CONTENTS

理论部分

第1章 Silverlight 概述 ······ 3

1.1 Silverlight 简介 ······ 3
    1.1.1 什么是 Silverlight ······ 3
    1.1.2 Silverlight 提供的功能 ······ 5
    1.1.3 Silverlight 发展史 ······ 5
    1.1.4 Silverlight 体系结构 ······ 9
    1.1.5 Silverlight 生命周期 ······ 12

1.2 Silverlight 开发 ······ 15
    1.2.1 获取工具 ······ 15
    1.2.2 安装 Visual Web Developer Express ······ 19
    1.2.3 安装 Silverlight 工具 ······ 19

1.3 创建第一个 Silverlight 应用程序 ······ 20

第2章 矢量绘图、画刷与着色 ······ 27

2.1 为 ASP.NET 赋予新的生命 ······ 27
    2.1.1 Silverlight 面世之前 ······ 27
    2.1.2 支持 Silverlight 的图形和动画 ······ 28

2.2 Silverlight 中的图形 ······ 28
    2.2.1 Shape ······ 28
    2.2.2 路径绘图 ······ 34
    2.2.3 几何绘图 ······ 37
    2.2.4 用 C# 绘制图形 ······ 41
    2.2.5 色彩概念 ······ 43
    2.2.6 画刷类型 ······ 44
    2.2.7 使用 C# 代码绘制画刷 ······ 49

第3章 图像与视觉特效 ······ 51

3.1 图像对象 ······ 51

3.1.1 图片拉伸属性 …… 52
3.1.2 图像画刷 …… 53
3.2 在 C# 中使用图像 …… 54
3.3 使用 BitmapImage 的下载事件 …… 55
3.3.1 使用 WriteableBitmap 绘制位图 …… 56
3.3.2 文本画刷应用 …… 57
3.4 透明特效 …… 58
3.5 透明遮罩 …… 59
3.6 裁剪特效 …… 61
3.7 RenderTransform 特效 …… 62
3.7.1 TranslateTransform 对象 …… 63
3.7.2 RotateTransform 对象 …… 64
3.7.3 ScaleTransform 对象 …… 65
3.7.4 SkewTransform 对象 …… 65
3.7.5 TransformGroup 对象 …… 66
3.7.6 MatrixTransform 对象 …… 67
3.7.7 在 C# 中应用变形对象 …… 69
3.7.8 Silverlight 3D Effects …… 71
3.7.9 关于 Element-To-Element Binding …… 74
3.8 Silverlight 3 Effect 特效 …… 74
3.8.1 BlurEffect …… 74
3.8.2 DropShadowEffect …… 77

第 4 章 动画与多媒体 …… 79

4.1 故事板和事件触发器 …… 79
4.2 Silverlight 线性插值动画 …… 81
4.2.1 DoubleAnimation 动画 …… 82
4.2.2 ColorAnimation 动画 …… 83
4.2.3 PointAnimation 动画 …… 85
4.3 Silverlight 关键帧动画 …… 86
4.3.1 DoubleAnimationUsingKeyFrames 动画 …… 87
4.3.2 ColorAnimationUsingKeyFrames 动画 …… 90
4.3.3 PointAnimationUsingKeyFrames 动画 …… 91
4.4 使用 C# 管理动画 …… 93
4.5 Silverlight 多媒体格式与通信协议 …… 95
4.5.1 MediaElement 支持与不支持的视频和音频格式 …… 95
4.5.2 MediaElement 的媒体播放机制 …… 97
4.6 MediaElement 对象 …… 97
4.7 视频拉伸模式 …… 98

4.8　MediaElement 状态管理 …… 99
4.9　缓冲进度和下载进度 …… 99
4.10　获取和控制播放位置 …… 101
4.11　视频画刷的应用 …… 101

第 5 章　Silverlight 与 HTML、JavaScript 三者交互 …… 103

5.1　Silverlight 对象模型与 DOM …… 103
5.2　获取 Silverlight 插件的错误信息 …… 105
5.3　在 Silverlight 中获取初始化参数和网页参数 …… 106
5.4　Silverlight 捕获浏览器信息 …… 107
5.5　Silverlight 操作 HTML 元素 …… 108
5.6　HTML 元素操作 Silverlight 对象 …… 110
　　5.6.1　使用 HttpUtility 类 …… 111
　　5.6.2　使用 Document.Cookies 读写 Cookie …… 113
　　5.6.3　使用 HtmlPage.Window 类 …… 116
5.7　Silverlight 调用 JavaScript …… 119
5.8　使用 JavaScript 调用 Silverlight …… 122

第 6 章　数据访问与 Silverlight 高级应用实例 …… 124

6.1　数据访问与远程通信 …… 124
　　6.1.1　WebClient …… 124
　　6.1.2　WebClient 与 XmlReader …… 125
　　6.1.3　WebRequest …… 129
　　6.1.4　实现跨域访问 …… 130
　　6.1.5　Silverlight 调用 WCF 服务 …… 131
6.2　文件打开对话框与文件上传 …… 135
6.3　使用保存文件对话框 …… 139
6.4　启用 Silverlight 应用程序库缓存 …… 140
6.5　应用控件截图功能 …… 141
6.6　Silverlight 3 Easing 动画集合 …… 143
6.7　使用墨迹控件 InkPresenter …… 144
6.8　使用 Silverlight 控件导航 …… 146

上 机 部 分

上机 1　Silverlight 概述 …… 155
　　第 1 阶段　指导 …… 155
　　　　指导 1　新建一个 Silverlight 程序 …… 155
　　　　指导 2　初识 Blend …… 156

第 2 阶段　练习 ………………………………………………………………………… 161
　　　练习　使用 Blend 工具制作简单动画 ……………………………………… 161

上机 2　矢量绘图、画刷与着色 ……………………………………………………… 162

第 1 阶段　指导 ………………………………………………………………………… 162
　　　指导 1　使用 Blend 或 VS2008 制作星光特效 …………………………… 162
　　　指导 2　使用 VS2008 制作 Silverlight 取色器 …………………………… 164
第 2 阶段　练习 ………………………………………………………………………… 167
　　　练习　使用绘图元素绘制销售统计图形 ……………………………………… 167

上机 3　图像与视觉特效 ……………………………………………………………… 168

第 1 阶段　指导 ………………………………………………………………………… 168
　　　指导 1　实现水中倒影效果 …………………………………………………… 168
　　　指导 2　运用 Silverlight 3D 特性制作三维空间 …………………………… 169
第 2 阶段　练习 ………………………………………………………………………… 175
　　　练习　运用 Silverlight 打造特效工具栏 …………………………………… 175

上机 4　动画与多媒体 ………………………………………………………………… 176

第 1 阶段　指导 ………………………………………………………………………… 176
　　　指导 1　结合动画与控件开发跑马灯图片浏览器 …………………………… 176
　　　指导 2　全功能视频播放器 …………………………………………………… 181
第 2 阶段　练习 ………………………………………………………………………… 186
　　　练习　制作 Silverlight 时钟效果 …………………………………………… 186

上机 5　Silverlight 与 HTML、JavaScript 三者交互 …………………………… 188

第 1 阶段　指导 ………………………………………………………………………… 188
　　　指导 1　创建一个 Silverlight 程序 ………………………………………… 188
　　　指导 2　保存 Cookie …………………………………………………………… 189
第 2 阶段　练习 ………………………………………………………………………… 190
　　　练习　读取 Cookie 文件中的用户名密码实现自动登录 …………………… 190

上机 6　数据访问与 Silverlight 高级应用实例 …………………………………… 192

第 1 阶段　指导 ………………………………………………………………………… 192
　　　指导 1　实现用户登录 ………………………………………………………… 192
　　　指导 2　实现员工管理的新增 ………………………………………………… 201
第 2 部分　练习 ………………………………………………………………………… 209
　　　练习　实现员工管理的删除和修改 …………………………………………… 209

理论部分

第1章

Silverlight 概述

学习目标

- ➤ 了解 Silverlight 概况
- ➤ 了解 Silverlight 开发工具
- ➤ 掌握 Silverlight 体系结构

1.1 Silverlight 简介

1.1.1 什么是 Silverlight

Silverlight 是微软公司在.NET Framework 平台上实现为 Web 和移动设备构建并显示下一代多媒体体验和丰富的交互式应用程序(RIA)的一种跨浏览器、跨平台的插件。Silverlight 技术是一种新的 Web 表现层技术,其跨平台的用户体验和可扩展的编程模型分别起到了统一服务器、Web 和桌面,统一托管代码和动态语言、声明性编程和传统编程以及 Windows Presentation Foundation (WPF)的功能,并通过结合音视频、动画、交互以及绚丽的用户界面为 Web 应用程序提供精彩的多媒体创意和丰富的交互式环境。

Silverlight 技术的前身称为 WPF/E(Windows Presentation Foundation Everywhere)技术,是利用跨浏览器 Web 技术进行设计,以实现多操作系统,甚至移动设备上无缝运行的一种基于声明性编程方式的 WPF,是一种新的 Web 呈现技术。

Silverlight 提供了 Windows Presentation Foundation (WPF)具有的一部分功能(如数据绑定、触发器、样式、路由事件、依赖项属性、可视化和逻辑树以及可冻结对象),并包含了可视状态管理器、Deep Zoom、数据网格控件等功能,因此可以将.NET Framework 部署经验转移到 Silverlight,生成易于部署和快速安装的 RIA。反之,也能够通过重复使用 XAML 将 Silverlight 应用程序移植到桌面。

因此从功能上说,Silverlight 隶属于 WPF 的子集,是传达下一代网页多媒体的交互性功能，并可与 Ajax 做高弹性的、程式化互动的、跨浏览器上的外挂。但二者都基于 XAML 的展示层基础，它们是互补的。与 Adobe 技术相比，Silverlight 犹如 Flash，WPF 犹如 AIR（前身为 Apollo），XAML 犹如 MXML。

Silverlight 有以下几个特点：

(1) 跨平台的用户体验：Silverlight 集成了多种现有 Web 技术和设备支持多种平台，使用户能够方便地进行代码重用，并通过不同平台无缝地连接到 Web。Silverlight 目前已经支持 Internet Explorer、Firefox 和 Safari 和谷歌浏览器，并可在 Microsoft Windows 和 Apple Mac OS X 上运行。

(2) 小巧方便：当用户遇到使用 Silverlight 开发的网页时，可以迅速地安装 Silverlight 插件，安装简单、体积小（约为 2MB）。

(3) 该解决方案集成了强大的图像及图层技术，支持任何尺寸图像的无缝整合，并提供适合广播的图层技术，可以在图像上添加按钮、标题或其他交互性内容。

(4) 丰富的内容功能：使用 Silverlight，可以添加包括视频、动画、文字、二维(2D)图像、三维(3D)图像和一些 Web 页面的可视化效果等丰富的内容，比单纯使用 HTML 带来了更丰富的用户体验。

(5) 可扩展的编程模型和协作工具：Silverlight 兼容大量其他标准和现有技术（包括 ASP.NET、AJAX 以及 .NET 3.5），支持 JavaScript、C#、VB、Ruby 以及 Python 等多种开发语言，使得开发者可以根据现有标准，或采用微软已成熟技术来开发基于 Web 的内容。而且 XAML 语言使得 Web 内容与桌面内容的开发语言一致，从而降低了开发费用。Silverlight 还为设计者和开发者提供大量的开发工具和开发环境支持（如 Visual Studio® 的 Web 开发支持包括 ASP.NET AJAX 在内的技术，Expression® Studio 使得在符合 W3C 标准的网站开发中可以使用 XHTML、XML、XSLT、CSS 以及 ASP.NET 等工具且能够创建可重用界面以提高创建 Silverlight 应用的效率）。

(6) 无须编译：Silverlight 基于 XAML 和 JavaScript，由浏览器解释执行，并以 DOM 形式公开它的元素树，内容能很好地被搜索引擎收录。

(7) 高质量、低成本的多媒体技术：Silverlight 内建的视频及动画广告解决方案灵活性很高，当传输广播类型的视频或动画广告时，不会影响视频的质量。它通过 Expression Media Encoder 及 Tarari 公司的内建平台，支持 15X 的快速视频编码及硬件加速，并且允许利用 WMV 标准从高清设备向移动设备提供高质量视频和音频。通过获得艾美奖的 Windows Media 技术，传输流量可降低 46%，并且与现有的 Windows Media 流量配置方案兼容。

(8) 结合数据、服务器和服务：XAML 与 ASP.NET AJAX 无缝集成，比单独使用 ASP.NET AJAX 提供了更丰富的表现能力。

(9) 支持内容接入保护技术：无论是在 Windows 平台还是 Mac 平台上，Silverlight 都支持多种商业模型，包括订阅、租用、付费浏览或预览等。

如图 1-1 所示为具有丰富图形和用户交互的 Silverlight 应用程序。

开发人员可以用多种方式创建 Silverlight 应用程序，可以使用 Silverlight 标记创建媒体和图形，并使用动态语言和托管代码来操作它们。Silverlight 还允许用户使用专业级别的工具，如使用 Visual Studio 进行编码和使用 Microsoft Expression Blend 进行布局和图形设计。

图 1-1　翻页屏幕快照

1.1.2　Silverlight 提供的功能

Silverlight 将多种技术组合到单个开发平台,可以允许用户根据需要选择合适的工具和编程语言。Silverlight 提供了下列功能:

(1) WPF 和 XAML。Silverlight 包含 Windows Presentation Foundation(WPF)技术的一个子集,从而大大扩展了浏览器中用于创建 UI 的元素。Silverlight 允许用户创建沉浸式图形、动画、媒体和其他丰富的客户端功能,使基于浏览器的 UI 远超单独使用 HTML 提供的效果。XAML 提供一个声明性标记语法用于创建元素。

(2) 对 JavaScript 的扩展。Silverlight 提供对通用浏览器脚本语言的扩展,可以控制浏览器 UI,包括使用 WPF 元素。

(3) 跨浏览器、跨平台支持。Silverlight 可以在所有通用浏览器(以及任意平台)上自如运行。用户可以设计和开发应用程序而不必担心使用何种浏览器或平台。

(4) 与现有应用程序集成。Silverlight 可以与现有 JavaScript 和 ASP.NET AJAX 代码无缝集成,以增强已具有的功能。

(5) 可以访问 .NET Framework 编程模型。可以使用诸如 IronPython 的动态语言以及诸如 C♯和 Visual Basic 的语言创建 Silverlight 应用程序。

(6) 工具支持。可以使用诸如 Visual Studio 和 Expression Blend 之类的开发工具快速创建 Silverlight 应用程序。

(7) 网络支持。Silverlight 包括对 TCP 上的 HTTP 的支持。可以连接到 WCF、SOAP 或 ASP.NET AJAX 服务并接收 XML、JSON 或 RSS 数据。

(8) LINQ。Silverlight 包括语言集成查询(LINQ),这种查询允许用户使用直观本机语法和 .NET Framework 语言中的强类型对象来编程进行数据访问。

1.1.3　Silverlight 发展史

1. Silverlight 1.0

2007 年 9 月 4 日,微软公司发布了 Silverlight 1.0 正式版,同时发布了新产品 Expression Encoder 1.0,Silverlight 1.0 的重点在于网页中的 RIA 体验。体积小巧的 Silverlight 1.0

运行时内置了解码器,可以在许多浏览器中播放 VC-1、WMV 等视频,以及 MP3、WMA 等音频文件,Silverlight 1.0 的口号是"增强您的 Web 体验"。

2. Silverlight 1.1

在 Silverlight 1.0 正式版本发布不久,微软公司推出 Silverlight 1.1 Alpha,它继承了 Silverlight 1.0 大部分特性,在 Silverlight 1.0 的基础上添加了托管代码的编程能力和对象模型,开发者可以使用 C♯或 VB 等托管语言来编写 Silverlight 程序,这是 Silverlight 技术的一个重大的变革,使得 Silverlight 技术进一步地向微软公司的.NET 框架特征靠拢。由于 Silverlight 1.1 Alpha 功能强大,可扩展性强,2007 年年底,微软公司宣布直接将 Silverlight 1.1 Alpha 更名为 Silverlight 2.0,这一声明来自微软公司开发人员部门的集团副总裁"Soma"Somasegar 的博客,他对这一变化的原因进行了说明(http://blogs.msdn.com/somasegar),新版本 Silverlight 将支持.NET,所以开发人员可以使用 C♯、VB、Ruby 和 Python 这些他们所期望的语言来编写 Silverlight 应用,相比 Silverlight 1.1 Alpha 而言,Silverlight 2.0 的特性得到了极大的提升。

3. Silverlight 2

2008 年 10 月 13 日,微软公司推出 Silverlight 2.0 的 RTW 版,并提供相应程式及档案的下载安装。

Silverlight 2.0 支持下列新功能特色,弥补了 Silverlight 1.0 所欠缺的功能。

(1) 具备一个.NET Framework 缩小版的基础类别函式库。

(2) 大量内建的 Silverlight 控制项:在 Silverlight 1.0 时,所有 UI 物件都必须通过 XAML 自行描述绘制,并缺乏许多内建的 Silverlight 向量控制项。针对这点,Silverlight 2.0 强化控制项方面的能力,内建许多向量控制项供开发人员直接使用。

(3) Skinning and Templating 外观样板的进阶支持:通过 Skinning and Templating 的支持,可以自定义控制项的外观与样板,可以迅速及动态地套用不同的外观。

(4) Deep Zoom:一个高解析度的影像缩放技术,能够在 Silverlight 进行深度的图片影像缩放。

(5) 广泛的网络 Networking 支持能力:举例来说支持 REST、WS*/SOAP、POX、RSS 及标准 HTTP 服务等网络技术呼叫,这部分对前端的 Silverlight 特别重要,通过网络程序才能存取后端 Server 的资料,以回传并显示在 UI 之上。

(6) 扩展的.NET Language 支持:Silverlight 2.0 不仅仅支持主流的 C♯及 VB 程式开发,亦进一步支持动态语言,例如 IronPython、IronRuby 等。

(7) Silverlight DRM 的支持:Silverlight 2.0 对于影音媒体内容的保护,是通过 DRM 技术来达成的,通过它可以提供 Content 内容保护。

(8) 改善服务端的延展性及扩展广告客户支持:Silverlight 2.0 针对串流传送资料方式、效率、下载播放方式再进一步强化改善其能力。

(9) 活跃的合作伙伴生态系统:微软公司全球知名的 Visual Studio Industry Partners 合作伙伴包括了 Component One LLC、Infragistics Inc 及 Telerik,提供了 Silverlight 2.0 的商业元件,可在 Visual Studio 2008 的环境中使用。

(10) 跨平台及跨浏览器支持:支持 Mac、Windows 及 Linux 等平台和 Firefox、Safari 与 Windows Internet Explorer 等浏览器。

4. Silverlight 3

2009年7月10日,微软公司正式发布 Silverlight 3.0。它具有如下优势:

1) 支持更多的媒体编码格式

在 Silverlight 3 中新增加的多媒体编码格式包括 H.264、AAC、MP4。Silverlight 技术从诞生以来,就一直把对多媒体尤其是视频的支持放在首要位置。这次对更多的编码格式提供支持,方便网站建设者更容易地发布、部署视频资料,只需要添加一行 XML 代码。

2) 利用 GPU 加速

随着显卡计算能力的加强,应用程序把越来越多的图形计算任务从 CPU 中拿出来交给 GPU 完成。然而,要利用到 GPU,对程序员而言通常意味着更多的编码任务,在 Silverlight 3 中,这个任务的复杂程度被大大简化,只需要在 XAML 中添加几行 XML 代码,就可以轻松享受 GPU 的超强计算能力。

3) 透视化 3D

透视化 3D,简单地说,就是把 2D 对象放到 3D 空间中去。与传统的 3D 把一个 3D 空间的对象投影到 2D 空间中不同,透视化 3D 意味着更高的性能,更友好的编程接口,同时能完成 80% 的 3D 任务。

4) 自定义特效

Silverlight 3 中引入了 shader 的概念,它是一个像素粒度的操作——每当 Silverlight 3 要显示一个像素时,它对 shader 说:"我要显示这个像素了,你是否要做些处理,实现某些特效?"下面举例说明 shader 的强大之处。在一个示例中,左边的图像是背景图像,右边的图像是前景图像,在 Silverlight 3 之前,尽管可以同时显示这两个图像,但是背景图像会被前景遮住(当然可以设置前景的透明度,但是这样整个前景图像的清晰度就下降了)。如果在显示前景图像时应用一个 shader,则可以把所有的黑色过滤掉。这样,在前景上过滤掉所有黑色背景的同时,还拥有了一个清晰的火焰。

目前,除了官方提供的阴影(shadow)和模糊(blur)shader 使用 GPU 加速,其他的 shader 仍需运行在 CPU 上。

5) 更多的控件支持

每一个新的 Silverlight 版本的发布,都伴随着很多新的控件问世。这个版本也不例外,新推出的控件有 DockPanel、Expander、Label、TreeView、ViewBox 等。在这里举一个"Save as…"(另存为)控件的例子。可能有的朋友会问,"另存为"作为一个耳熟能详的控件,为什么要等到 Silverlight 3 才提供?答案是出于对安全性的考虑,即 Silverlight 为了保护用户的安全,对本地文件的读写有很大的限制(否则,如果服务器端可以随意读写用户的本地文件,那么将很容易做出一个钓鱼网站)。在 Silverlight 3 中,开发人员可以创建一个 SaveFileDialog 实例,但是当用户选定本地文件时,他只能得到这个文件的 stream,而不是这个文件的路径。这样的设计避免了提供打开任意路径本地文件的功能,使 Silverlight 运行在一个更为安全的环境中。

6) 本地消息传递(local messaging)

Silverlight 是浏览器的一个插件,在同一时间可能会有多个实例。例如,当多个浏览器同时访问包含 Silverlight 的网页时,会有多个 Silverlight 实例同时运行,本地消息传递允许这些不同的 Silverlight 控件实例之间互相通信。

7) 在浏览器外运行 Silverlight(out of browser)

Silverlight 3 支持把一个 Silverlight 页面安装到本地,用户可以像使用桌面程序一样离线使用这个程序,并且可以通过右键菜单卸载这个程序。

5. Silverlight 4

2010 年 4 月 16 日,微软公司正式发布 Silverlight 4.0。新特性使 Silverlight 开发的 Web 应用程序将会越来越接近桌面应用程序的便携性,在这里列出新版本中最吸引人的 10 个新特性。

(1) 网络摄像机和麦克风允许为聊天或客户服务应用程序共享视频和音频。

(2) 视频和音频本地记录功能可以直接捕捉原始视频,扩大应用场景,如视频会议。

(3) 可以在应用程序中使用复制、粘贴、拖曳操作。

(4) 支持常规桌面操作模型,如右键菜单。

(5) 多播网络支持,减少企业流媒体广播成本,这样企业可以广泛开展视频培训,视频会议,并可以和现有 Windows 媒体服务器基础设施无缝集成。

(6) 读写用户"我的文档""我的音乐""我的图片"和"我的视频"文件夹下的文件。

(7) 运行其他桌面程序,如 Office。例如,启动 Outlook 发送一封电子邮件时,发送一个报告到 Word,或发送数据到 Excel。

(8) 通过 COM 访问设备或其他系统功能,如访问 USB 安全卡阅读器。

(9) 广泛的打印支持,并支持虚拟打印视图,完全独立于屏幕内容。

(10) .NET 通用运行时(CLR)让相同的代码不用修改即可运行在桌面上,又可运行在 Silverlight 中。

当然还有其他新特性。最令人感兴趣的还是如何利用 Silverlight 编写如传统桌面应用程序般好用的 Web 应用程序。同时要告诉大家的是 Silverlight 4 的速度是 Silverlight 3 的 2 倍,还支持 3D 加速,今后使用 Silverlight 4 开发的游戏将会有更好的体验。

6. Silverlight 5

2011 年 4 月 13 日,微软正式发布 Silverlight 5 Beta 版本。Silverlight 5 带来了对 IE 9 的完美支持,其在 Silverlight 4 的基础上新增了 40 多个新功能,完善了媒体支持并提供了更加丰富的用户界面,主要新功能包括:

(1) 通过使用 GPU 为低功耗设备渲染高清视频提供了 H.264 硬件解码功能。

(2) "TrickPlay"允许以不同的速度播放视频,且支持快进、后退,以及音高修正,也就是说在快速播放视频的同时,以正常速度播放声音。

(3) 改进了电源管理,在播放视频时阻止屏幕保护程序的启动,允许计算机在视频停止播放后休眠。

(4) 远程控制支持,允许用户远程控制媒体播放。

(5) 增强的数字版权管理允许在 DRM 媒体源间无缝切换。

作为微软的下一代企业应用程序解决方案,Silverlight 5 还完善了在企业应用开发方面的特性,其完善的特性如下:

(1) 在用户界面内可以实现更流畅的动画效果,改进了字体渲染清晰度,支持 Postscript 矢量打印。

(2) 支持 64 位操作系统;IE 9 无窗口模式的硬件加速。

（3）XAML 解析性能改进，加速启动和运行性能。

（4）数据绑定和 MVVM 增强。

（5）GPU 加速的 3D API 支持高级数据可视化呈现和富用户界面和体验。

（6）测试工具——增加 Visual Studio 2010 的自动化 UI 测试支持。

（7）脱离浏览器模式——Silverlight 5 应用已经可以创建并管理子窗口，受信任的脱离浏览器应用可以使用 P/Invoke 功能等。

1.1.4 Silverlight 体系结构

本节介绍 Microsoft Silverlight 的基本结构和组件。Silverlight 不仅是一个漂亮的画布，可用来向最终用户显示丰富的交互式 Web 内容和媒体内容，它还是一个功能强大的轻量平台，可用来开发可移植、跨平台的网络应用程序，以便与来自许多源的数据和服务集成。此外，使用 Silverlight 生成的用户界面与传统 Web 应用程序相比，前者显著增强了典型的最终用户体验。

虽然，Silverlight 作为客户端运行时环境看起来简单轻便，但 Silverlight 开发平台集成了大量功能和复杂技术，并使它们可供开发人员访问。若要创建基于 Silverlight 的高效应用程序，开发人员需要具备该平台结构的使用知识。

1. Silverlight 平台

Silverlight 平台作为一个整体，由两个主要部分以及一个安装程序和更新组件组成，如表 1-1 所示。

表 1-1 Silverlight 平台组成部分

组 件	描 述
核心表示层框架	面向 UI 和用户交互的组件和服务（包括用户输入、用于 Web 应用程序的轻量型 UI 控件、媒体播放、数字版权管理和数据绑定）以及表示层功能（包括矢量图形、文本、动画和图像），还包括用于指定布局的可扩展应用程序标记语言（XAML）
.NET Framework for Silverlight	.NET Framework 中包含组件和库的一个子集，其中包括数据集成、可扩展 Windows 控件、网络、基类库、垃圾回收和公共语言运行时（CLR）。.NET Framework for Silverlight 的某些部分是通过应用程序部署的。这些"Silverlight 库"是未包括在 Silverlight 运行时但在 Silverlight SDK 中提供的程序集。在应用程序中使用 Silverlight 库时，它们会与应用程序打包在一起，并下载到浏览器中。这些库包括新的 UI 控件、XLINQ、整合（RSS/Atom）、XML 序列化和动态语言运行时（DLR）
安装程序和更新程序	是一个安装和更新控件，可简化安装操作过程

图 1-2 演示了 Silverlight 结构的这些组件以及相关组件和服务。

由 Silverlight 平台中包括的工具、技术和服务组成的组合集具有特殊价值，即它们使开发人员能够更方便地创建丰富的交互式网络应用程序。尽管使用目前的 Web 工具和技术无疑也能生成此类应用程序，但开发人员会遇到很多技术难题，其中包括不兼容的平台、不同的文件格式和协议、以不同方式呈现网页和处理脚本的各种 Web 浏览器。在一个系统和

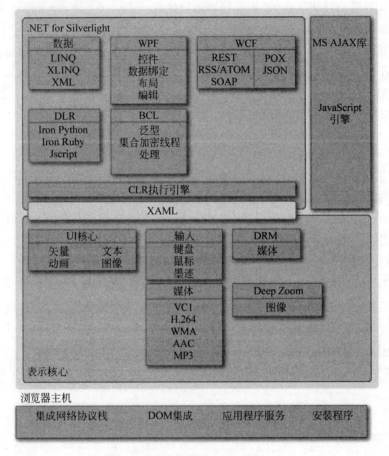

图 1-2 Silverlight 结构

浏览器中能够正常运行的丰富 Web 应用程序在另一个系统或浏览器中的运行效果会完全不同，也可能会失败。要构建具有以下优点的应用程序，使用目前的大批工具、协议和技术可能需要付出巨大的努力并且成本极高：

（1）跨浏览器和平台创建相同的用户体验，使应用程序的外观和执行效果保持一致。

（2）使用熟悉的.NET Framework 类和功能将来自多个网络位置的数据和服务集成到一个应用程序中。

（3）引人注目且易于访问的富媒体用户界面（UI）。

（4）Silverlight 使开发人员更容易生成此类应用程序，因为它克服了当前技术的许多不兼容性，并且在一个平台内提供了可用于创建跨平台的丰富集成应用程序的工具。

2．核心表示层组件

表 1-2 介绍了 Silverlight 平台的核心表示层功能。

表 1-2　Silverlight 平台的核心表示层功能

功　　能	描　　述
输入	处理来自硬件设备（例如键盘和鼠标、绘图设备和其他输入设备）的输入
UI 呈现	呈现矢量和位图图形、动画以及文本

续表

功　能	描　述
媒体	具有播放和管理各种类型音频和视频文件(如.WMP和.MP3文件)的功能
Deep Zoom	使用户能够放大高分辨率图像和围绕该图像进行平移
控件	支持可通过应用样式和模板来自定义的可扩展控件
布局	可以动态定位UI元素
数据绑定	可以链接数据对象和UI元素
DRM	可以对媒体资产启用数字版权管理
XAML	为XAML标记提供分析器

开发人员可以通过使用XAML指定表示层详细信息来与此表示层框架进行交互。XAML是.NET Framework和表示层之间的主要交互点。开发人员可以使用托管代码以编程方式操作表示层。

3. .NET Framework for Silverlight

表1-3介绍了.NET Framework for Silverlight功能列表中的部分功能。

表1-3 .NET Framework for Silverlight 功能列表

功　能	描　述
数据	支持语言集成查询(LINQ)和LINQ to XML功能,这些功能简化了集成和使用不同源数据的过程。还支持使用XML和序列化类来处理数据
基类库	一组.NET Framework库,这些库提供了基本编程功能,如字符串处理、正则表达式、输入/输出、反射、集合和全球化
Windows Communication Foundation(WCF)	提供的功能可简化对远程服务和数据的访问。其中包含浏览器对象、HTTP请求和响应对象、对跨域HTTP请求的支持、对RSS/Atom整合源的支持以及对JSON、POX和SOAP服务的支持
CLR(公共语言运行时)	提供内存管理、垃圾回收、类型安全检查和异常处理
WPF(Windows Presentation Foundation)控件	提供了一组丰富的控件,其中包含Button、CheckBox、HyperlinkButton、ListBox、RadioButton、ScrollViewer
DLR(动态语言运行时)	支持动态编译和执行脚本语言(如JavaScript和IronPython),以编写基于Silverlight的应用程序。包括一个可插接式模型,用来添加对Silverlight所使用的其他语言的支持

.NET Framework for Silverlight是完整版.NET Framework的一个子集。它为以前不支持的应用程序类型(如Internet应用程序)提供了面向对象的可靠应用程序的基本开发。

开发人员可以通过使用C♯和Visual Basic编写托管代码来与.NET Framework for Silverlight层进行交互。.NET Framework开发人员还可以通过在Visual Studio或Microsoft Expression Blend中进行创作访问表示层。

4. 附加Silverlight编程功能

Silverlight提供了多个可帮助开发人员创建丰富的交互式应用程序的附加功能,包括表1-4中介绍的功能。

表 1-4 Silverlight 附加功能

功 能	描 述
独立存储	提供从 Silverlight 客户端到本地计算机的文件系统的安全访问。可以将本地存储和数据缓存与特定用户隔离
异步编程	当应用程序被释放以便进行用户交互时,后台工作线程会执行编程任务
文件管理	提供一个安全的"打开文件"对话框,以简化创建安全文件上载的过程
HTML-托管代码交互	.NET Framework 程序员可以直接操作网页 HTML DOM 中的 UI 元素。Web 开发人员也可以使用 JavaScript 直接调用托管代码,以及访问可编写脚本的对象、属性、事件和方法
序列化	支持将 CLR 类型序列化为 JSON 和 XML
打包	提供用于创建.xap 包的 Application 类和生成工具。.xap 包中包含要运行 Silverlight 插件控件所需的应用程序和入口点
XML 库	XmlReader 和 XmlWriter 类简化了使用 Web 服务中的 XML 数据的过程。开发人员借助 XLinq 功能可使用.NET Framework 编程语言直接查询 XML 数据

1.1.5 Silverlight 生命周期

1. Silverlight 生命周期

一个 Silverlight 应用程序,从开始请求到完全载入,一般经过 6 个步骤:

(1) 用户请求 HTML 页面,HTML 页面中含有 Silverlight 的入口。

(2) 浏览器下载 Silverlight 插件和对应的 XAP 文件。

(3) Silverlight 插件开始工作,读取 XAP 文件里的 AppManifast.xml 文件来载入需要用到的 Assemblies。

(4) Silverlight 插件创建一个 App 类的实例。

(5) Application 的默认构造函数引发 Startup 事件。

(6) Silverlight 应用程序处理 Startup 事件。直到 Silverlight 应用遇到异常或用户关闭页面,Silverlight 应用程序结束。

2. 应用程序入口

由 Silverlight 的生命周期看到,Silverlight 插件下载好了以后创建的第一个类就应该是 App 类,因此,App 类的默认构造函数成为 Silverlight 的应用程序入口。下面打开 app.xaml.cs 来看 App 类的默认构造函数:

```
public partial class App : Application
{
    public App()
    {
        this.Startup += this.Application_Startup;
        this.Exit += this.Application_Exit;
        this.UnhandledException +=
          this.Application_UnhandledException;
        InitializeComponent();
    }
}
```

可见，App 什么都没有做，只是添加了 3 个事件的处理程序，分别对应了 Silverlight 的启动、退出和出现异常。而真正意义上的代码，发生在应用程序的 Startup 事件中：

```
private void Application_Startup(object sender, StartupEventArgs e)
{
    this.RootVisual = new MainPage();
}
```

这里虽然只有一行代码，但是它调用了主窗口的构造函数，并且，将生成的对象赋给了 App 对象的 RootVisual 属性。这样，Silverlight 应用程序就被引入到用户的代码片段中。

3．应用程序事件

在应用程序载入的过程中可以看到，App 类的构造函数中注册了 3 个应用程序事件的处理程序。App 类中用很大篇幅处理这 3 个重要的应用程序事件。

4．Startup

Silverlight 应用程序启动时触发，前节中已经涉及其处理程序，它用于生成启动页面类的实例并且赋值给 Application.Current.RootVisual。另外，可以通过 e 获取初始化的参数。方法如下。

首先，把需要的参数添加到 HTML 中：

```
<object data="data:application/x-silverlight-2,"
  type="application/x-silverlight-2" width="100%" height="100%">
    <param name="source" value="ClientBin/ApplicationModel.xap"/>
    <param name="onError" value="onSilverlightError"/>
    <param name="background" value="white"/>
    <param name="minRuntimeVersion" value="3.0.40818.0"/>
    <param name="autoUpgrade" value="true"/>
    <param name="initParams" value="initWords1=Hello,initWords2=World"/>
    <a href="http://go.microsoft.com/fwlink/?LinkID=149156&v=
    3.0.40818.0" style="text-decoration:none">
        <img src="http://go.microsoft.com/fwlink/?LinkId=161376"
          alt="Get Microsoft Silverlight" style="border-style:none"/>
    </a>
</object>
```

这个参数必须以"initParams"命名，参数的键与值之间以等号分隔；多个参数之间用逗号分隔。然后，就可以通过 e.InitParams 对初始化参数进行访问了，例如：

```
private void Application_Startup(object sender, StartupEventArgs e)
{
    this.RootVisual = new MainPage();
    if (e.InitParams.ContainsKey("initWords2"))
    {
        TextBlock tb = new TextBlock();
        tb.Text = e.InitParams["initWords2"].ToString();
        ((MainPage)this.RootVisual).LayoutRoot.Children.Add(tb);
```

```
            }
        }
```

取出参数列表中的第二个参数,把它赋给一个 TextBlock 并且添加到 MainPage 的 Grid 中。

5. Exit

Exit 事件在 Silverlight 退出时,会被触发。这一事件可以给程序一个最后的机会来保存用户相关的数据,如记录当前用户的偏好等。

6. UnhandledException

顾名思义,UnhandledException 会在有未处理的异常被系统捕获到时触发。Silverlight 生成了一些默认的处理程序:

```
private void Application_UnhandledException(object sender,
 ApplicationUnhandledExceptionEventArgs e)
{
    //如果应用程序是在调试器外运行,则使用浏览器的异常机制报告该异常。在 IE 上,
    //将在状态栏中用一个黄色警报图标来显示异常,而 Firefox 则会显示一个脚本错误。
    if (!System.Diagnostics.Debugger.IsAttached)
    {
        //注意:这使应用程序可以在已引发异常但尚未处理该异常的情况下继续运行。对于
        //    生产应用程序,此错误处理应替换为向网站报告错误并停止应用程序。
        e.Handled = true;
        Deployment.Current.Dispatcher.BeginInvoke(delegate {
            ReportErrorToDOM(e);
        });
    }
}
```

e.Handled 用来表明这个异常是不是已经被处理了。当 Debug 时,e.Handled 会被设置为 true,并且通过一个匿名方法异步地调用 ReportErrorToDOM 方法,而这个方法则把异常信息显示出来:

```
private void ReportErrorToDOM(ApplicationUnhandledExceptionEventArgs e){
    try{
        string errorMsg = e.ExceptionObject.Message +
          e.ExceptionObject.StackTrace;
        errorMsg = errorMsg.Replace('"', '\'')
           .Replace("\r\n", @"\n");
        System.Windows.Browser.HtmlPage.Window.Eval("throw
          new Error(\"Unhandled Error in Silverlight 2 Application "
          + errorMsg + "\");");
    }catch (Exception){
    }
}
```

1.2 Silverlight 开发

1.2.1 获取工具

为了开始 Microsoft Silverlight 开发,最简单的方式就是使用 Microsoft Web 平台安装程序(Web PI)。这个简单的应用程序可帮助用户安装和配置许多东西,包括工具、服务器、数据库、编程 API 和应用程序。

使用 Web 平台安装程序的操作如下。

Web PI 可以从 http://www.microsoft.com/web 免费下载。安装并启动 Web PI 后,会看到如图 1-3 所示的窗口,此处选择"Web 平台"标签。

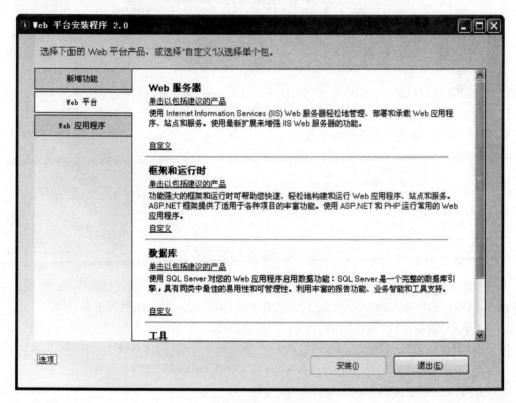

图 1-3　选择"Web 平台"标签

第一个区域"Web 服务器"(如图 1-4 所示)允许安装"Internet 信息服务"(IIS) Web 服务器和配置大量选项,它们涉及应用程序开发,常用 HTTP 功能,与之前版本的兼容性,部署和发布,健康和诊断,管理、性能和安全性等。在"Web 服务器"下方有一个"单击以删除包括的产品"链接。单击该链接,随后会出现一个绿色"√",如图 1-4 所示。

第二个区域是"框架和运行时"(图 1-5)。可在这里安装和配置 Microsoft Web 平台的各种开发功能,包括 ASP.NET、ASP.NET MVC 和 PHP。如果"框架和运行时"旁边没有出现"√",单击"单击以删除包括的产品"链接。

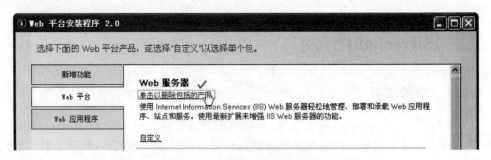

图 1-4　单击"单击以删除包括的产品"链接的结果

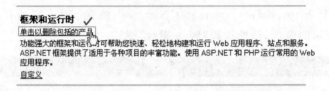

图 1-5　"框架和运行时"选项

"数据库"区域（图 1-6）允许安装 SQL Server Express 引擎。可利用它向 Web 应用程序添加数据库。以后通过本书学习开发时，会用到这样的数据库。由于要使用的一些工具不在默认列表中，所以需要进行如下操作。

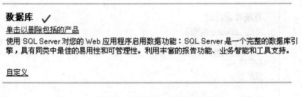

图 1-6　"数据库"选项

（1）如果"数据库"（图 1-6）旁边没有绿色"√"，单击"单击以删除包括的产品"链接。如果这个链接不可用，则表明已经安装了建议的工具，所以可以安全地跳过当前步骤。

（2）单击"数据库"区域底部的"自定义"链接。随后会显示一个新屏幕，允许自定义要安装 SQL Server 的哪些部分。此时，会看到已经勾选了"SQL Server Express 2008 R2"，但没有勾选"SQL Server 2008 R2 Management Studio Express"，如图 1-7 所示。

（3）勾选"SQL Server 2008 R2 Management Studio Express"，单击"返回 Web 平台"。

最后一个区域是"工具"，它允许安装和配置用于 ASP.NET、Silverlight、JavaScript 等的开发工具。和本节之前讨论的一样，如果"工具"旁边没有出现绿色"√"，就会显示"单击以删除包括的产品"链接。请单击"自定义"，勾选所有复选框，然后单击"返回 Web 平台"，如图 1-8 所示。

至此，已经准备好开始安装了。单击窗口底部的"安装"按钮，Web PI 会显示要安装的所有项目的列表，要求接受许可条款。单击屏幕底部的"我接受"按钮继续，如图 1-9 所示。

由于选择安装 SQL Server Express，所以会出现图 1-10 所示的另一个对话框。它询问如何设置安全性。有两个选择。第一个是使用"集成 Windows 身份验证"，也就是用

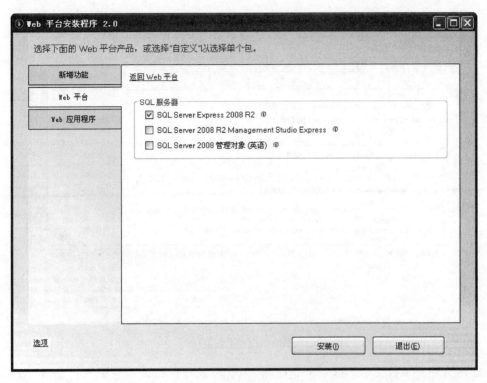

图 1-7 数据库自定义

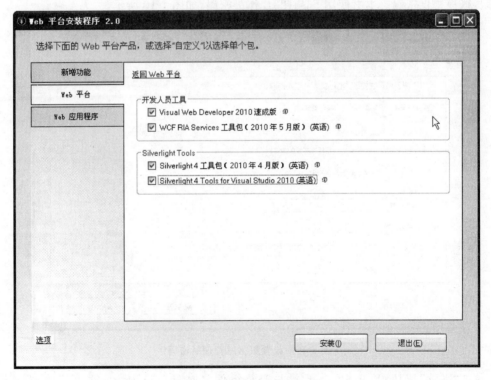

图 1-8 "返回 Web 平台"选项

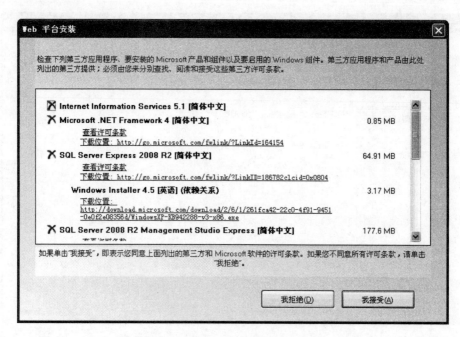

图 1-9　接受安装

Windows 账户登录数据库。第二个是使用"混合模式身份验证",它既支持集成 Windows 身份验证,也允许 SQL Server 拥有它自己的登录系统。请选择"混合模式身份验证",并为管理员账户提供密码。本书使用 Sasa123!作为密码,如图 1-10 所示。

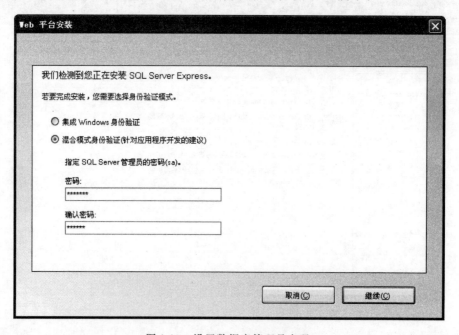

图 1-10　设置数据库管理员密码

单击"继续"按钮,Web PI 会下载并安装组件。如图 1-11 所示,这可能需要一定的时间。

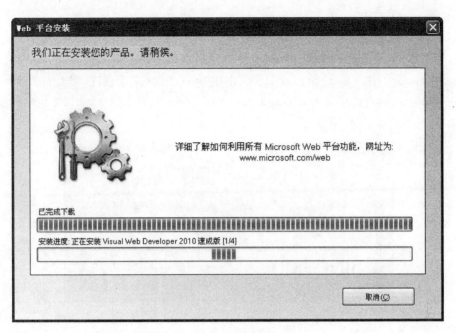

图 1-11　安装进度条

完成之后，Microsoft Web 平台、数据库和进行开发所需的全部依赖项便安装好了。注意，在某些版本的 Windows 7 中，可能出现一条消息，提醒 SQL Server 2008 Management Studio Express 在 Windows 7 中有一些已知的问题。如果出现该消息，直接单击"运行"按钮忽略。

1.2.2　安装 Visual Web Developer Express

Microsoft Visual Studio 套装软件价格不菲，但可以考虑使用它的免费版本，也就是所谓的 Express 版或学习版。

可选择以下 Express 版：
- Microsoft Visual Basic 2010 Express
- Microsoft Visual C# 2010 Express
- Microsoft Visual C++ 2010 Express
- Microsoft Visual Web Developer 2010 Express

1.2.3　安装 Silverlight 工具

Silverlight 是一个正在快速演变的平台，这里讨论的一些细节等你读到本书时可能已经发生了变化！无论如何，获取最新信息最理想的地方是 http://www.silverlight.net。下载 Silverlight 工具的最佳地点是 http://www.silverlight.net/getstarted。

所有 Silverlight 工具都包括以下组件：
- Silverlight 运行时
- Silverlight Tools for Visual Studio（提供了与 VS 的集成，并提供了相应的模板，方便用户开始构建 Silverlight 应用程序）

- Silverlight SDK
- 一些附加的控件
- WCF RIA Services 包（便于构建一些商业应用程序）

除此之外，为了在 Visual Studio 中顺利创建 Silverlight 应用程序，需要安装 Silverlight 4 for Developers，请直接访问 http://go.microsoft.com/fwlink/? LinkId＝146060 下载并安装，如图 1-12 所示。

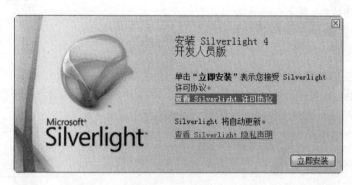

图 1-12　安装 Silverlight 4 for Developers

1.3　创建第一个 Silverlight 应用程序

安装好一切后，即可创建第一个应用程序。本节将创建一个简单的 Silverlight 应用程序，并对它进行解析。

1. 构建简单的 Silverlight 应用程序

打开"开始"菜单，启动 Visual Web Developer 2010 Express，如图 1-13 所示。

图 1-13　启动 Visual Web Developer 2010 Express

为了创建一个新的 Silverlight 应用程序，请选择"文件"→"新建项目"，如图 1-14 所示。

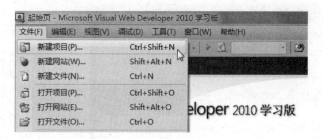

图 1-14　新建项目

随后会出现如图 1-15 所示的"新建项目"对话框，其中列出了已安装的模板。模板是应用程序的"骨架"。选定一个模板后，Visual Web Developer 会创建这种应用程序所需的一切基本文件。模板按编程语言组织，每个模板都有 Visual Basic 和 Visual C# 版本。打开 Visual C# 文件夹，会看到多种不同的应用程序类型，其中包括 Windows、Web、Cloud 和 Silverlight。

图 1-15　"新建项目"对话框

选择"Silverlight 应用程序"模板。在"名称"文本框中输入 SbSCh1_1，然后单击"确定"按钮。随后，Visual Web Developer 会开始创建 Silverlight 应用程序。Silverlight 应用程序需要在 Web 上运行，所以需要一个网站，以便在其中运行。Visual Web Developer 能自动创建该网站。

随后将出现如图 1-16 所示的"新建 Silverlight 应用程序"对话框，询问是否想在新网站中承载 Silverlight 应用程序。确定已勾选了该复选框。然后，Silverlight 会创建和 Silverlight 项目同名的 Web 项目，但使用后缀 .Web。其他选项保持默认值不变，然后单击"确定"按钮。

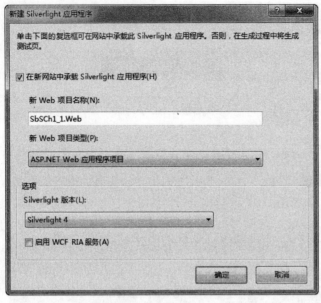

图 1-16 选择承载程序

图 1-17 目录结构

Visual Web Developer 会创建一个新的解决方案。在一个解决方案中,可以对不同的项目进行组织。解决方案将包含两个项目:Silverlight 应用程序和用于容纳 Silverlight 应用程序的网站。可在图 1-17 所示的解决方案资源管理器中清晰地看到这一点。

Silverlight 应用程序使用"可扩展应用程序标记语言"(XAML)文件描述用户界面(UI)。在 MainPage.xaml 文件中包含默认 UI。双击该文件,在设计器中打开 UI。屏幕左侧竖直显示了一组标签,分别是"工具箱""文档大纲"和"数据源"。打开"工具箱"标签页,如图 1-18 所示。

单击工具箱右上角的图钉按钮,使工具箱固定在屏幕上。单击"通用 Silverlight 控件"区域,随后会看到简单 Silverlight 控件的一个列表,如图 1-19 所示。

图 1-18 工具箱

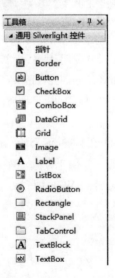

图 1-19 固定工具箱

第一个应用程序将使用两个 Label 控件、一个 TextBox 控件和一个 Button 控件。为了添加第一个 Label 控件,双击工具箱中的 Label 控件。随后会发生两件事情。

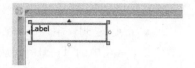

图 1-20　添加 Label

(1) Silverlight 在设计平面添加 Label 的一个可视形式(图 1-20)。

(2) 将 Label 的标记代码添加到 XAML 中(图 1-21)。

```
<UserControl x:Class="SbSCh1_1.MainPage"
    xmlns="http://schemas.microsoft.com/winfx/2006/xaml/presentation"
    xmlns:x="http://schemas.microsoft.com/winfx/2006/xaml"
    xmlns:d="http://schemas.microsoft.com/expression/blend/2008"
    xmlns:mc="http://schemas.openxmlformats.org/markup-compatibility/2006"
    mc:Ignorable="d"
    d:DesignHeight="300" d:DesignWidth="400" xmlns:sdk="http://schemas.microsoft.com/winfx/2006/xaml/presentation/sdk">

    <Grid x:Name="LayoutRoot" Background="White">
        <sdk:Label Height="28" HorizontalAlignment="Left" Margin="10,10,0,0" Name="label1" VerticalAlignment="Top" Width="120" />
    </Grid>
</UserControl>
```

图 1-21　添加 Label 的标记

现在需要编辑 Label,使其不显示默认的"Label"字样。有两个方法可以执行这个操作。可以使用"属性"窗口更改 Content 属性的值。"属性"窗口默认显示在屏幕右下角,按 F4 键可以打开或关闭它。在 Content 属性右侧的文本框中输入"你的名字是什么?",如图 1-22 所示。

图 1-22　输入 Content 属性

除此之外,还可以直接编辑 XAML。为此,在 sdk:Label 标记中添加一个名为 Content 的属性,将它的值设为"你的名字是什么?"。注意,如果使用"属性"窗口来设置,会自动在 XAML 中添加相应的属性,如图 1-23 所示。

```
<Grid x:Name="LayoutRoot" Background="White">
    <sdk:Label Height="28" HorizontalAlignment="Left" Content="你的名字是什么?" Margin="10,10,0,0"
        Name="label1" VerticalAlignment="Top" Width="120" />
</Grid>
```

图 1-23　添加控件并设置属性

配置好第一个 Silverlight 控件后,还要添加其他控件。

2. 配置更多 Silverlight 控件

双击工具箱中的 TextBox,从而在设计器中添加一个 TextBox 控件。注意,TextBox 被添加到刚才创建的 Label 的正下方。可以使用鼠标拖动 TextBox,把它定位到 Label 的右侧,如图 1-24 所示。

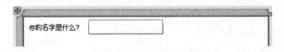

图 1-24　添加控件并设置属性

重复这个过程添加一个 Button 控件。把它拖放到 TextBox 右侧,将 Content 属性更改为"Go",而不是默认的"Button"。注意,虽然 Button 是不同的控件类型,但添加和配置其 Content 属性的方式和 Label 控件是一样的。

要添加的最后一个 UI 元素是另一个 Label。双击工具箱中的 Label 添加一个新控件。新控件会自动定位到原始 Label 的下方。保持这个位置不变,但要用鼠标拖动 Label 的右侧,使它变得更宽,如图 1-25 所示。Label 右侧会显示一个小圆点。将鼠标移到这里,指针会变成一个左右箭头。在这种情况下,按住鼠标左键不放,向右拖动鼠标即可。

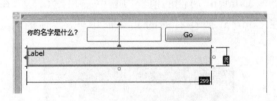

图 1-25　改名控件的宽高

注意一下 XAML 代码。检查每个控件的 Name 属性。默认情况下,Visual Web Developer 在对控件命名时采用的是控件类型加一个编号的方式。第一个 Label 控件名为 label1,第二个 Label 名为 label2,以此类推。在当前这个应用程序中,存在着名为 label1、label2、textBox1 和 button1 的控件。在实际应用中,最好是为控件指定更有意义的名称,此时保持默认值不变。

接着,添加少许代码使应用程序能真正起效果。双击 Go 按钮,切换到代码窗口,表明进入代码编辑模式。

Visual Web Developer 会自动创建一个名为 button1_Click 的存根函数。要在其中写入代码,以便在用户按下该按钮时执行某程序。

3. 为按钮添加功能

在 button1_Click 函数中输入单词 label2。在输入字符的过程中,会自动弹出"智能感知"菜单。

"智能感知"根据.NET Framework 中已安装的类以及应用程序中的控件实例来推测接下来要输入什么内容。由于目前唯一以"Lab"开头的只有 Label 类(在.NET Framework 中)和当前应用程序中的 label1 和 label2 控件,所以"智能感知"会将选择范围缩减至这 3 个选项。可坚持自己输入 label2,也可从菜单中选择 label2,如图 1-26 所示。

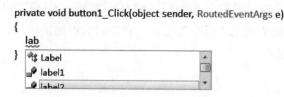

图 1-26 "智能感知"功能

为了访问 .NET Framework 中的一个控件的属性，要使用句点语法。例如，label2 的 Content 属性是通过 label2.Content 来访问的。在 label2 之后输入一个句点符号之后，"智能感知"会再次介入，列出该控件可以访问的所有属性。从中选择 Content 属性（或直接输入）。

完成整行代码，如下所示：

```
private void button1_Click(object sender, RoutedEventArgs e)
{
    label2.Content = "Hello World, " + textBox1.Text;
}
```

不管用户在文本框中输入什么，输入的文本都会和"Hello World,"连接，结果字符串被赋给 label2 标签控件的 Content 属性。所以，一旦单击 Go 按钮，就会在 label2 中显示文本"Hello World,"加上文本框中输入的名字。

按 F5 键运行应用程序。随后会启动默认浏览器，并运行新的 Silverlight 应用程序。请在文本框中输入一个名字，并单击 Go 按钮，label2 将显示文本"Hello World,"加这个名字，如图 1-27 所示。

图 1-27 运行 Silverlight 应用程序

至此，创建、设计、编码、编译、部署和运行了第一个 Silverlight 应用程序。

总 结

本章讲解了 Silverlight 的基本概念以及 Silverlight 体系结构，然后介绍了安装 Silverlight 插件的方法。Silverlight 运行时，插件是使用用户浏览器运行 Silverlight 应用程序的先决条件，预览和开发 Silverlight 应用程序之前都必须安装相应版本的 Silverlight 插件。学习本章之后,你是否感觉到做出绚丽的网页很简单?体验到 Silverlight 应用程序的魅力之后,你是不是也跃跃欲试了呢?那么在本章之后会详细讲解 Silverlight 的特效、动画以及与 ASP.NET 集成。

1. 简述你对 Silverlight 的理解。
2. 简述 Silverlight 的优势。

第2章 矢量绘图、画刷与着色

学习目标

- 掌握三大绘图(形状绘图、路径绘图、几何绘图)
- 掌握通过 C# 托管代码绘制图形
- 掌握两种画刷(线性渐变画刷和放射性画刷)的使用

2.1 为 ASP.NET 赋予新的生命

本节介绍在当前技术中正确定位 Silverlight 及其能力,讨论 Silverlight 可能为现有的 ASP.NET 应用程序带来的一些好处。

2.1.1 Silverlight 面世之前

有人认为在 Silverlight 面世之前,Web 就是令人厌烦的、静态的世界,是非常幼稚的,事实并非如此,Web 中还存在大量的图形和动画。在 Silverlight 之前,多种技术可以用于提供具有较大影响的图像、动画和丰富的内容,而最值得注意的是 Flash 和 Java applet。

Flash 为大多数人打开了编写多功能 Web UI 之门,它提供了一个相当容易的编程模型,并且该模型主要集中在图形和动画处理上。Java applet 提供了同样的功能。但是,直接跳到 Java 的学习曲线对于很多人而言困难重重。

除了这两项技术之外,XHTML+SMIL 语言也提供了在 Web 上对图形、媒体和动画的支持,但是该语言比较少用。

同步多媒体集成语言(Synchronized Multimedia Integration Language,SMIL)是一种基于 XML 的语言。它可以用于提供定时和动画,以及内嵌多媒体内容。XHTML+SMIL 是当前为了在 HTML 页面中的 SMIL 而提供的 W3C 规范。最初,微软公司、Macromedia 和 Compaq/DEC 就集成 HTML 和 SMIL 给 W3C 提交了一个提案,该语言称为 HTML+TIME。XHTML+SMIL 构建在该语言之上,并且添加了另外的一些功能,因此 XHTML+

SMIL 变成了当前完成该任务的标准。

2.1.2 支持 Silverlight 的图形和动画

Silverlight 解决了丰富的功能总是要以陡峭的学习曲线为代价的问题。正如前面已经提到的,通过 XAML 和.NET 隐藏代码实现 UI 布局和过程代码的分离,意味着设计人员可以快速地掌握该技术,或者通过手工完成,或者使用开发工具的 Expression 套件。那么留给.NET 开发人员(人员多得难以计数)的任务就是编写需要的代码。

Silverlight 可以创建"打破模式"的 UI/各个组成部分完全可以是自定义的,而不用依赖内置控件/形状来构建另外一个普通的站点或插件。本章将深入研究复杂 2D 图形,在屏幕上渲染所有可能的形状,而不是编写一行过程代码。其具有完整功能的动画 API 可以为内容带来生命,例如可以在精确的时间内方便地在屏幕上移动多个控件,同时平稳地改变颜色。

简而言之,Silverlight 可以通过提供集成设计时所需要的丰富图形和动画以及易于设计者使用的环境,为已有的 ASP.NET 站点赋予新的生命。

2.2　Silverlight 中的图形

使用 Silverlight 来实施 UI 开发将会遇到很多问题,而且这些问题仅仅依赖于标准内置控件的 UI 开发中是不曾遇到的。也许设计团队希望输入控件看上去就像他们的公司标记——你可能知道的一个图标,具有圆角边框和填充内容。或者如果图形/视频能够更好地适应到特定形状的容器,而不是使用一个普通矩形的话,那么它们看上去可能会好一些。

不管出于什么原因,这样的需求都将在 Silverlight 中得到满足。

2.2.1　Shape

System.Windows.Shapes.Shape 类是 Silverlight 中 6 个基本形状类型 Line、Ellipse、Rectangle、Polyline、Polygon 和 Path 的基类。Shape 继承自 FrameworkElement,因此也就继承自 UIElement,这使得该类具有接受输入并获取输入焦点的能力,而且还可以参与布局——简而言之,它可以在屏幕上实施渲染。继承关系如图 2-1 所示。

Shape 类在其基类基础上添加了多个混合属性,包括:通过其 Fill 属性提供的在其内进行绘图的能力;通过其 Stretch 属性提供的以某种方式对其进行拉伸,以充满其容器空间的能力;通过和 StrokeThickness 属性提供的指定如何渲染形状轮廓的能力。

Shape 是一个抽象类,这意味着它不能直接被实例化。为了使用该类,必须从该类派生出新的类。

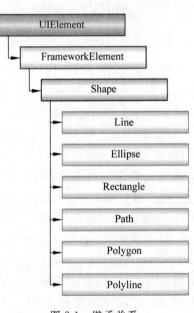

图 2-1　继承关系

1. Line 对象

线条(Line)是 Shape 类中的一个最基本的绘图元素。Line 顾名思义,在两点之间画出一条直线,需要指定起始点(X1、Y1)和终结点(X2、Y2)的坐标。例如:

```
<Line X1="50" Y1="50" X2="150" Y2="150" Stroke="Black"
   StrokeThickness="3"></Line>
```

运行结果如图 2-2 所示。

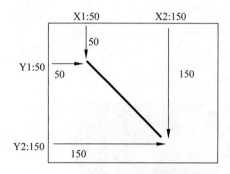

图 2-2　线条元素的绘图结果

上面代码演示了如何使用线条(Line),X1、Y1 确定了线条的起始点,X2、Y2 确定了线条的终点。

下面利用线条绘制一个组合图形,绘制一条折线,每一条线的终点是第二条线的开始点,代码如下:

```
<!--第一条线-->
<Line X1="50" Y1="50" X2="150" Y2="150" Stroke="Black"
   StrokeThickness="3"></Line>
<!--第二条线-->
<Line X1="150" Y1="150" X2="500" Y2="50" Stroke="Black"
   StrokeThickness="3"></Line>
```

运行结果如图 2-3 所示。

Line 线条本身还支持渐变的填充,下面例子演示将线条的颜色填充为渐变的。代码如下:

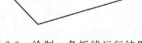

图 2-3　绘制一条折线运行结果

```
<!--水平直线填充-->
    <Line X1="10" Y1="50" X2="300" Y2="50" StrokeThickness="10" Margin="50,0,0,0">
        <Line.Stroke>
            <LinearGradientBrush>
                <GradientStop Color="Gold" />
                <GradientStop Color="White" Offset="1" />
```

```
            </LinearGradientBrush>
        </Line.Stroke>
    </Line>
    <!-- 垂直直线填充 -->
    <Line StrokeThickness = "10" X1 = "350" Y1 = "50" X2 = "350" Y2 = "200" Margin = "50,0,0,0">
        <Line.Stroke>
            <LinearGradientBrush>
                <GradientStop Color = "Blue" />
                <GradientStop Color = "White" Offset = "1" />
            </LinearGradientBrush>
        </Line.Stroke>
    </Line>
```

运行结果如图 2-4 所示。

图 2-4　为 Line 元素加入渐变填充的效果

Stroke、StrokeThickness 是很多 Shape 元素的公共属性值,用来定义元素边框的颜色和宽度值。

2．Ellipse 对象

Ellipse 即椭圆形,主要是通过声明 width 和 height 属性来创建,如果设置 width 和 height 相等,画出来将是圆形,其主要的属性还有 Stroke(对边框线填充)、StrokeThickness(边框线的宽度)、Fill(对图形进行填充)。例如:

```
<Canvas Background = "#CDFCAE">
    <Ellipse Canvas.Top = "20" Canvas.Left = "40"
        Width = "160" Height = "80" Fill = "#FF9900"
        Stroke = "Black" StrokeThickness = "3">
    </Ellipse>

    <Ellipse Canvas.Top = "20" Canvas.Left = "260"
        Width = "180" Height = "100">
        <Ellipse.Fill>
            <RadialGradientBrush GradientOrigin = "0.5,0.5" Center = "0.5,0.5"
                RadiusX = "0.5" RadiusY = "0.5">
                <GradientStop Color = "#0099FF" Offset = "0" />
                <GradientStop Color = "#FF0000" Offset = "0.25" />
                <GradientStop Color = "#FCF903" Offset = "0.75" />
                <GradientStop Color = "#3E9B01" Offset = "1" />
            </RadialGradientBrush>
```

```
            </Ellipse.Fill>
        </Ellipse>

        <Ellipse Canvas.Top = "120" Canvas.Left = "160"
            Width = "100" Height = "100" Fill = "#FF9900"
            Stroke = "#000000" StrokeThickness = "2">
        </Ellipse>
</Canvas>
```

运行结果如图 2-5 所示，显示 3 个椭圆形。

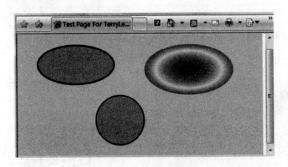

图 2-5　绘制椭圆

3. Rectangle 对象

Rectangle 是矩形，在创建矩形时并不需要提供每个角的坐标值，只需要使用 width 和 height 属性即可，如果设置 width 和 height 相等则为正方形，其主要的属性还有 Stroke（对边框线填充）、StrokeThickness（边框线的宽度）、Fill（对图形进行填充），同时还可以通过 RadiusX 和 RadiusY 设置圆角效果，例如：

```
<Canvas Background = "#CDFCAE">
    <Rectangle Canvas.Top = "20" Canvas.Left = "40"
        Width = "160" Height = "80" Fill = "#FF9900"
        Stroke = "Black" StrokeThickness = "3">
    </Rectangle>

    <Rectangle Canvas.Top = "20" Canvas.Left = "260"
        Width = "180" Height = "100">
        <Rectangle.Fill>
            <RadialGradientBrush GradientOrigin = "0.5,0.5" Center = "0.5,0.5"
                RadiusX = "0.5" RadiusY = "0.5">
                <GradientStop Color = "#0099FF" Offset = "0" />
                <GradientStop Color = "#FF0000" Offset = "0.25" />
                <GradientStop Color = "#FCF903" Offset = "0.75" />
                <GradientStop Color = "#3E9B01" Offset = "1" />
            </RadialGradientBrush>
        </Rectangle.Fill>
    </Rectangle>
```

```
          <Rectangle Canvas.Top = "120" Canvas.Left = "120"
              Width = "100" Height = "100"
              Stroke = "#000000" StrokeThickness = "2" RadiusX = "15" RadiusY = "15">
              <Rectangle.Fill>
                  <LinearGradientBrush StartPoint = "0,1">
                      <GradientStop Color = "#FFFFFF" Offset = "0.0" />
                      <GradientStop Color = "#FF9900" Offset = "1.0" />
                  </LinearGradientBrush>
              </Rectangle.Fill>
          </Rectangle>
      </Canvas>
```

运行结果如图 2-6 所示，其中有两个加上了渐变效果。

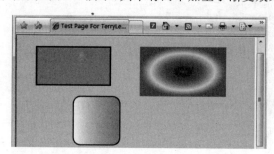

图 2-6 绘制矩形

4. Polyline 和 Polygon 对象

Polyline（多线段）和 Polygon（多边形）是 Silverlight 中简单但功能强大的绘图对象。它允许用户声明多个 Points 属性值，然后根据 Points 属性的坐标先后顺序来绘制图形，并且支持线条内部的颜色填充。

首先介绍 Polyline，它比 Line 元素更加强大，下面例子利用 Polyline 来绘制线条。

```
<Polyline Stroke = "Black" Points = "0,0 50,50 300,0 400,100"></Polyline>
```

运行结果如图 2-7 所示。

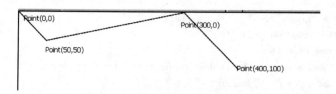

图 2-7 用 Polyline 来绘制线条

代码中 Points = "0,0 50,50 300,0 400,100" 总共分成了 4 对坐标，然后根据 4 个点来绘制图形。

下面的例子来说明了 Polyline（多线段）和 Polygon（多边形）的用法和区别。

```
<!-- 三组坐标组成的 Polyline 开放图形 -->
<Polyline Stroke = "Blue" StrokeThickness = "6" Points = "50,50 150,
200 300,50" Fill = "Gold"></Polyline>
<!-- 三组坐标组成的 Polygon 封闭图形 -->
<Polygon Stroke = "Blue" StrokeThickness = "6" Points = "400,50 650,
200 800,50" Fill = "Gold"></Polygon>
```

运行结果如图 2-8 所示。

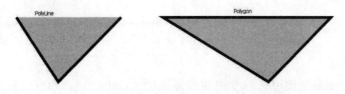

图 2-8　用 Polyline 和 Polygon 元素绘制图形

上面例子分别利用 Polyline（多线段）和 Polygon（多边形）来绘制图形，除坐标不一样外，其他的均基本相同并填充了相同的颜色，唯一区别是，Polygon 绘制出来的图形是一个封闭的图形。

Polyline（多线段）和 Polygon（多边形）均具有 FillRule（填充规则），使用 FillRule 可以声明线条内的填充规则，例如：

```
<StackPanel x:Name = "LayoutRoot" Background = "White" Orientation = "Horizontal">
    <!-- EvenOdd 确定一个点是否位于填充区域内的规则,具体方法是从该点沿任意方向画一条
无限长的射线。
    然后计算该射线在给定形状中因交叉而形成的路径段数。如果此数目为奇数,则该点在内部;
如果是偶数,则该点在外部。
    Nonzero 确定一个点是否位于路径填充区域内的规则,具体方法是从该点沿任意方向画一条无
限长的射线,然后检查形状段与该射线的交点。
    从零开始计数,每当线段从左向右穿过该射线时加 1,而每当路径段从右向左穿过该射线时
减 1。
    计算交点的数目后,如果结果为零,则说明该点在路径外部。否则,说明该点位于路径内部。
    -->

    <Polygon Stroke = "Blue" StrokeThickness = "2" Fill = "Orange" Margin = "120,0,60,0"
FillRule = "EvenOdd"
    Points = "15,200 68,70 110,200 0,125 134,125"></Polygon>

    <Polygon Stroke = "Blue" StrokeThickness = "2" Fill = "Red" Margin = "50,0,0,0" FillRule =
"Nonzero"
    Points = "15,200 68,70 110,200 0,125 134,125"></Polygon>
</StackPanel>
```

运行结果如图 2-9 所示。

图 2-9 使用 Polygon 绘制并使用不同的 FillRule 属性值来填充

2.2.2 路径绘图

路径(Path)是一种比较特殊的、用于描述比较复杂图形的元素,它支持从简单到复杂的任意图形的绘制。

1. 路径标记语法

使用 Path 绘图需要使用它的专用路径标记语法(mini-language)。它是一种由路径指令组成的语法,使用 mini-language 可以产生任何形状的 2D 图形。下面是路径标记语法(mini-language)所支持的指令。

1) FillRule

指定该路径使用 EvenOdd 还是 NonZero 填充规则值:F0 指定 EvenOdd 填充规则;F1 指定 Nonzero 填充规则。

如果省略此命令,则路径使用默认行为,即 EvenOdd。如果指定此命令,则必须将其置于最前面。

2) FigureDescription

图形由移动命令、绘制命令和可选的关闭命令组成。

```
MoveCommandDrawCommands[ CloseCommand ]
```

3) MoveCommand

MoveCommand 用来指定图形起点的移动命令,可以用 M 或 m 表示,M 表示绝对值,m 表示相对于上一点的偏移量。

4) DrawCommand

DrawCommand 是一个或多个描绘图形内容的绘制命令,属于一个指令的集合,用来描述外形轮廓的内容,包含 Silverlight 中大量的直线和曲线绘图指令。

5) CloseCommand

CloseCommand 是可选的关闭命令,用于关闭图形。用来闭合整个 Path 并在当前结束点与开始点之间画一条弦,用 z 表示。

Data 是 Path 的重要属性,可以用 Data 来产生不同形状的几何图形,使用时只需要将组合好的指令直接赋给 Data,Path 会根据 Data 中的指令来完成绘图指令。示例代码如下:

```
< Canvas x:Name = "LayoutRoot" Background = "White">
    < Path Stroke = "Blue" Fill = "Gold" < strong > Data = "M 50,200 L150,200 100,50z"</ strong > />
</Canvas >
```

运行结果如图 2-10 所示。

MoveCommand、DrawCommand、CloseCommand 这些指令集组成了 mini-language，如图 2-11 所示。

图 2-11 中 M 50,200 是 MoveCommand 移动指令，表示起始坐标是 50,200；L150,200 是 DrawCommand 直线指令，L 表示线条类型的终点坐标是 100,50；100,50z 表示闭合坐标是 100,50，z 是 CloseCommand 关闭指令，用来封闭整个图形的轮廓。

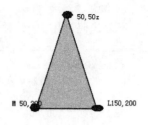

图 2-10 运行结果

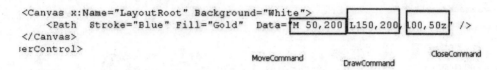

图 2-11 mini-language 指令集

2．绘图指令

DrawCommand 指令集合提供了十分丰富的 2D 形状。包括直线——Line（L）、Horizontal Line（H）和 Vertical Line（V），曲线——Cubic Bezier curve（C）、Quadratic Bezier curve（Q）、Smooth Cubic Bezier curve（S）、Smooth Quadratic Bezier curve（T）。

表 2-1 所示为 DrawCommand 指令集合支持的直线和曲线指令。

表 2-1　DrawCommand 指令集合支持的直线和曲线指令

指令名称	语法	描述
直线	L 或 l	在当前点和指定的终点之间画一条直线
水平线	H 或 h	在当前点和指定的 x 坐标之间画一条水平线。 格式：H 或 h
垂直线	V 或 v	在当前点和指定的 y 坐标之间画一条垂直线。 格式：V 或 v
三次贝塞尔曲线	C 或 c	通过控制点 1 和控制点 2，在当前点和指定的终点之间画一条三次贝塞尔曲线。 格式：C＋控制点 1＋控制点 2＋结束点或 c＋控制点 1＋控制点 2＋结束点
二次贝塞尔曲线	Q 或 q	通过指定控制点在当前点和指定的终点之间画一条二次贝塞尔曲线。 格式：Q＋控制点＋结束点或 q＋控制点＋结束点
平滑贝塞尔曲线	S 或 s	在当前点和指定的终点之间画一条平滑贝塞尔曲线。 格式：S＋控制点＋结束点或 s＋控制点＋结束点
平滑二次贝塞尔曲线	T 或 t	在当前点和指定的终点之间画一条平滑二次贝塞尔曲线
椭圆弧	A 或 a	在当前点和指定的终点之间画一个椭圆弧。 格式：(A 或 a)＋尺寸＋圆弧旋转角度＋大弧度形标记＋正负方向标记＋弧度终点

注：当绘制贝塞尔曲线时，需要提供 3 个坐标点，其中前两个坐标点是控制点，第三点是贝塞尔曲线的结束点。

3. 绘制直线和曲线

下面代码利用 Path 对象绘制直线和曲线并进行填充。

```
< StackPanel x:Name = "LayoutRoot" Background = "White" Orientation = "Horizontal">
    <! -- 绘制贝塞尔曲线 -->
    < Path Stroke = "Blue" StrokeThickness = "3" Fill = "Gold" Margin = "25,50,50,0" Data = "M 10,100 C 100,300 300, - 200 300,100"/>
    <! -- 绘制三角形 -->
    < Path Stroke = "Blue" StrokeThickness = "3" Fill = "Gold" Data = "M 0,200 L 200,200 100,50z" />
</StackPanel >
```

运行结果如图 2-12 所示。

示例代码中第一个图形是一个曲线,这个曲线的 Data 属性使用了绘图指令"C",这里的"C"代表了三次贝塞尔曲线(Cubic Bezier Curve)。100,300 是控制点 1,300,-200 是控制点 2,300,100 是控制结束点。

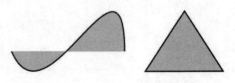

图 2-12　利用 Path 对象绘图

第二个图形是一个三角形,三角形的绘制由 0,200 开始,经过 200,200,最后到 100,50 坐标结束。

在使用绘制指令的过程中,指令的名称均以简写字母表示,例如,Line 简写为 L,Horizontal Line 简写为 H,Vertical Line 简写为 V。Data 属性包含的语法可以由一个或一个以上的绘图指令组成,指令字母对于大小写是敏感的。例如:

```
< StackPanel x:Name = "LayoutRoot" Background = "White" Orientation = "Horizontal">
    <! -- 大写 V 指令 -->
    < Path Stroke = "Green" StrokeThickness = "3" Data = "M 0,50 H 200 V 50 l 50,50" />
    <! -- 小写 v 指令 -->
    < Path Stroke = "Green" StrokeThickness = "3" Data = "M 0,50 H 200 v 50 l 50,50" />
</StackPanel >
```

运行结果见图 2-13。

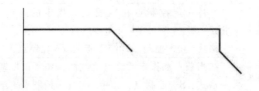

图 2-13　Vertical Line 指令和 Horizontal Line 指令绘图结果

上面例子中,分别利用了 H、V(v)、l 三个指令集合,大写 V 与小写 v 的效果显然区别很大,第一个 Path 的起始点是 H 50,50,所以 V 50 产生的垂直线条长度就为 0。但第二个 Path 使用的是小写 v,而小写 v 是相对于上一点坐标点作为起始点,那么它就产生了一条长 50 的垂直线条。

Plotline、Polygon 和 Path 三个绘图对象，Plotline、Polygon 比较简单，而 Path 则相对比较灵活。

补充：有时可能会遇到这种问题，即在编码过程中无论怎么编写，Path 对象都编译不出来，而且生成时会报错。这是因为项目中没有添加 Path 控件，此时需要在工具中右击选择"选择项"，在 Silverlight 组件中找到 Path 控件，勾选该控件后单击"确定"按钮即可，如图 2-14 所示。

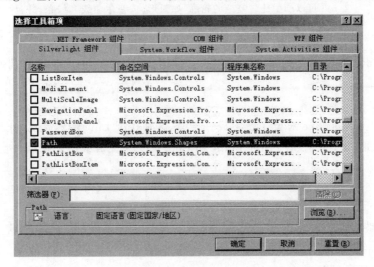

图 2-14　添加 Path 对象

2.2.3　几何绘图

在 Silverlight 的绘图能力之中，除了前面介绍的 Shape 类以外，Silverlight 还提供了一个几何绘图类，它可以提供比类更灵活的绘图对象。

1. Geometry 和 Shape 对象

Geometry（几何绘图）对象中包括 LineGeometry（几何线条）、RectangleGeometry（几何矩形）、EllipesGeometry（几何椭圆图形）、GeometryGroup（几何组合）、PathGeometry（几何路径），它可以描述任何几何的 2D 形状，Geometry 可以绘制简单的图形到复杂的几何路径。

从绘图来看 Geometry 类和 Shape 类似乎都是绘制 2D 图形，但是这两个类有着重要的区别，Geometry 类和 Shape 类之间并没有直接的联系，Geometry 类（几何绘图）相对于 Shape 类（形状绘图）更加轻量级，绘图效率更高于 Shape。

2. Geometry 和 Path 对象

LineGeometry（几何线条）、RectangleGeometry（几何矩形）、EllipesGeometry（几何椭圆图形）、GeometryGroup（几何组合）、PathGeometry（几何路径）都是由 Geometry 继承而来的。Geometry 集合的成员如表 2-2 所示。

表 2-2　Geometry 集合的成员

元素	描述
LineGeometry	代表一条直线对象，Line 对象的等价物
RectangleGeometry	代表一个矩形对象，Rectangle 对象的等价物

续表

元 素	描 述
EllipseGeometry	代表一个椭圆形对象，Ellipse 对象的等价物
GeometryGroup	把一定数量的 Geometry 集合对象添加到一个单独的 Path 上，使它们能够组合起来
PathGeometry	可以产生比较复杂的轮廓，如弧形、贝塞尔曲线和线条，并且可以控制线条是否封闭

事实上 Path 还可以作为一个容器，可以容纳任何 Geometry 形状的几何图形包含在 Path.Data 内。

3. LineGeometry 对象

类似于 Shape 的 Line 对象用来生成一条线，区别在于 Line 用的是 X 和 Y 坐标来生成线条，而 LineGeometry 是利用 StartPoint 和 EndPoint 来完成线条的绘制。代码如下：

```
<LineGeometry StartPoint = "0,0" EndPoint = "100,500" />
```

RectangleGeometry（几何矩形）、EllipesGeometry（几何椭圆图形）类似于 Shape 中的 Rectangle 和 Ellipse，这里不做过多描述。

4. GeometryGroup 对象

有时需要将某些图形组合起来，GeometryGroup 就具备这个功能，例如：

```
<StackPanel x:Name = "LayoutRoot" Orientation = "Horizontal" Background = "AliceBlue">
    <Path Fill = "Cyan" Stroke = "Black" StrokeThickness = "3">
        <Path.Data>
            <!-- GeometryGroup 组合 -->
            <GeometryGroup FillRule = "EvenOdd">
                <RectangleGeometry RadiusX = "2" RadiusY = "2" Rect = "80,50 200,100">
                </RectangleGeometry>
                <EllipseGeometry Center = "300,100" RadiusX = "80" RadiusY = "60">
                </EllipseGeometry>
            </GeometryGroup>
        </Path.Data>
    </Path>
    <Path Fill = "Cyan" Stroke = "Black" StrokeThickness = "3">
        <Path.Data>
            <!-- GeometryGroup 组合 -->
            <GeometryGroup FillRule = "Nonzero">
                <RectangleGeometry RadiusX = "2" RadiusY = "2" Rect = "80,50 200,100">
                </RectangleGeometry>
                <EllipseGeometry Center = "300,100" RadiusX = "80" RadiusY = "60">
                </EllipseGeometry>
            </GeometryGroup>
        </Path.Data>
    </Path>
</StackPanel>
```

运行结果如图 2-15 所示。

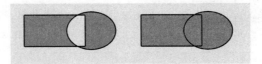

图 2-15　GeometryGroup 绘制组合图形

在两个图形交叉时,可以使用 Geometry 的 FillRule 属性定义组合图形的填充规则。FillRule 属性有两个枚举值(EvenOdd 和 Nonzero)。

5. PathGeometry 对象

PathGeometry 是 Geometry 中最灵活的,它可以绘制任意的 2D 几何图形。例如:

```
<Path Stroke = "Black" StrokeThickness = "1">
    <Path.Data>
            <!-- 指定 Path.Data 的填充是 PathGeometry -->
            <PathGeometry>
                <!-- 定义 PathGeometry 的 Figures -->
                <PathGeometry.Figures>
                    <PathFigureCollection>
                        <!-- PathFigure 表示几何图形的一个子部分、一系列单独连接的二维几何线段。
                        IsClosed:获取或设置一个值,该值指示是否连接该图形的第一条线段和最后一条线段。 -->
                        <PathFigure IsClosed = "True" StartPoint = "50,100">
                            <PathFigure.Segments>
                                <BezierSegment Point1 = "100,0" Point2 = "200,200" Point3 = "300,100"/>
                                <LineSegment Point = "400,100" />
                                <ArcSegment Size = "50,50" RotationAngle = "45" IsLargeArc = "False" SweepDirection = "Clockwise" Point = "200,100"/>
                            </PathFigure.Segments>
                        </PathFigure>
                    </PathFigureCollection>
                </PathGeometry.Figures>
            </PathGeometry>
    </Path.Data>
</Path>
```

运行结果如图 2-16 所示。

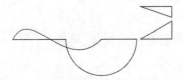

图 2-16　PathGeometry 绘图

为简化上面 XAML，wpf 提供了路径语法解析器，代码如下：

```
< Path Stroke = "Black" StrokeThickness = "1"
       Data = "M 10,100 L 100,100 100,50 Z M 10,10 100,10 100,40 Z" />
```

6. LineSegment 对象

利用 LineSegment 对象创建直线对象，代码如下：

```
< Path Stroke = "DarkCyan" StrokeThickness = "3">
        < Path.Fill >
            < LinearGradientBrush >
                < GradientStop Color = "Orange"/>
                < GradientStop Color = "White" Offset = "1"/>
            </LinearGradientBrush >
        </Path.Fill >
        < Path.Data >
            < PathGeometry >
                <! -- 指明是闭线条并且指定起始位置 -->
                < PathFigure IsClosed = "True" StartPoint = "50,100">

                    < LineSegment Point = "200,200" />
                    < LineSegment Point = "200,150" />
                    < LineSegment Point = "400,150" />
                    < LineSegment Point = "400,50" />
                    < LineSegment Point = "200,50" />
                    < LineSegment Point = "200,0" />
                </PathFigure >
            </PathGeometry >
        </Path.Data >
    </Path >
```

运行结果如图 2-17 所示。

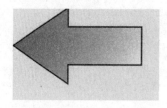

图 2-17　LineSegment 绘图

7. ArcSegment 对象

利用 ArcSegment 对象来绘制弧线元素，代码如下：

```
< Path Stroke = "DarkCyan" StrokeThickness = "3">
        < Path.Data >
            < PathGeometry >
```

```
            <!-- ArcSegment 指定弧的起始点 -->
            <PathFigure IsClosed = "False" StartPoint = "50,50">
                <!-- ArcSegment 声明第一条弧的结束点和弧度 -->
                <ArcSegment Size = "280,280" Point = "400,50" />
                <!-- ArcSegment 声明第二条弧的结束点和弧度 -->
                <ArcSegment Size = "90,280" Point = "550,150" />
                <ArcSegment Size = "50,50" Point = "600,50" />
            </PathFigure>
        </PathGeometry>
    </Path.Data>
</Path>
```

运行结果如图 2-18 所示。

8. BezierSegment 对象

利用 BezierSegment 对象绘制贝塞尔曲线,贝塞尔曲线是由比较复杂的数学公式产生的。它用来计算两个控制点之间如何确定一条曲线的轮廓。示例代码如下:

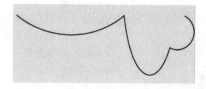

图 2-18 利用 ArcSegment 绘制弧形

```
<!-- 开始绘制贝塞尔曲线 -->
<Path Stroke = "DarkCyan" Fill = "YellowGreen" StrokeThickness = "5">
    <Path.Data>
        <PathGeometry>
            <!-- 声明贝塞尔曲线 -->
            <PathFigure StartPoint = "10,10">
                <BezierSegment Point1 = "130,30" Point2 = "40,140" Point3 = "150,150"/>
            </PathFigure>
        </PathGeometry>
    </Path.Data>
</Path>
```

运行结果如图 2-19 所示。

2.2.4 用 C# 绘制图形

在 Silverlight 中利用 C# 绘制图形比较简单,经常使用的两种方法是①直接创建对象,然后添加到页面容器中;②先创建 XAML 对象,然后利用 XamlReader.Load 方法加载到容器中。下面举例说明。

图 2-19 利用 BezierSegment 绘制贝塞尔曲线

第一种,直接创建对象,然后添加到页面容器中,代码如下:

```
public void DrawPolyLine()
{
```

```csharp
        //创建 Polyline
        Polyline polyline = new Polyline();
        //创建坐标集合
        PointCollection points = new PointCollection();
        points.Add(new Point(50,300));
        points.Add(new Point(50,50));
        points.Add(new Point(200,300));
        polyline.Points = points;
        //填充背景色与线条颜色
        polyline.Fill = new SolidColorBrush(Colors.Orange);
        polyline.Stroke = new SolidColorBrush(Colors.Black);
        polyline.StrokeThickness = 3;

        //设置位置
        Canvas.SetTop(polyline, 50);
        Canvas.SetLeft(polyline, 50);
        //向容器添加控件
        LayoutRoot.Children.Add(polyline);
}
public void DrawPolygon()
{
        //创建 Polygon
        Polygon polygon = new Polygon();
        //创建坐标集合
        PointCollection points = new PointCollection();
        points.Add(new Point(50, 300));
        points.Add(new Point(50, 50));
        points.Add(new Point(200, 300));
        polygon.Points = points;
        //填充背景色与线条颜色
        polygon.Fill = new SolidColorBrush(Colors.Orange);
        polygon.Stroke = new SolidColorBrush(Colors.Black);
        polygon.StrokeThickness = 3;
        //设置位置
        Canvas.SetTop(polygon, 50);
        Canvas.SetLeft(polygon, 300);
        //向容器添加控件
        LayoutRoot.Children.Add(polygon);
}
```

运行结果如图 2-20 所示。

图 2-20 用 C# 绘图第一种方法

第二种，先创建 XAML 对象，然后利用 XamlReader.Load 方法加载到容器中，代码如下：

```
public void DrawPathWithXML()
{
    string xaml = "< Path ";
    //引用
    xaml += " xmlns = \"http://schemas.microsoft.com/client/2007\"";
    //创建属性
    xaml += " Stroke = \"Blue\" StrokeThickness = \"3\" ";
    xaml += string.Format(" Data = \"{0}\" />", "M 10,100 Q100,200 300,100");
    //创建路径对象
    Path path = new Path();
    path = (Path)XamlReader.Load(xaml);
    LayoutRoot.Children.Add(path);
}
```

运行结果如图 2-21 所示。

图 2-21　用 C#绘图第二种方法

2.2.5　色彩概念

前面学习了在 Silverlight 中简单地画图，现在学习为这些图形着色。

在学习 Silverlight 的各种画刷对象之前，有必要先介绍一些计算机色值方面的概念。Silverlight 中的图形颜色一般由两种形式来描述。

1. 颜色名称

如 Black、Red、Blue、White、Yellow 等，这些颜色都是由比较标准的色值构成，直接使用名称即可得到需要的颜色，Silverlight 目前提供了 141 种颜色名称。

2. RGB 值

在实际应用中，当需要一个颜色名称中不包含的特殊颜色时，需要采用指定颜色值的方式来获取某一种颜色，Silverlight 的 Brush 元素直接支持 RGB 颜色填充，其中 R、G、B 分别代表红、绿、蓝 3 种单色，所以 RGB 也称为 RGB 三原色。

RGB 这 3 种颜色是合成任何一种颜色的基础，在 HTML 中，一个 RGB 颜色值由 3 个两位的十六进制数组成，例如 #FF0000 为标准的红色，这里的 #FF0000 等同于颜色名称里的 Red，之所以 #FF0000 代表红色，是因为红色达到了最高值 FF，而 FF 在十进制中代表 255，随后的 0000 分别代表绿色和蓝色，因为绿色值和蓝色值均为 00，所以，这里只呈现出标准的红色。

除了上面提到的 RGB 色值以外，Silverlight 还给提供了 Alpha 值，也就是对象的透明度，透明度值固定放在 RGB 颜色值的前面，所以 Silverlight 中的一组颜色值包含 4 个部分，即 ARGB，其中 A 代表 Alpha 透明度，那么一个标准的红色值可以写成 #FFFF0000，当 A

为 FF 时表示不透明，相反 A 为 00 时对象会完全透明，变得不可见，例如：

```
<Canvas Width="400" Height="320">
    <Ellipse Width="100" Height="100" Fill="#88FF0000" Canvas.Top="60" Canvas.Left="160"></Ellipse>
    <Ellipse Width="100" Height="100" Fill="#8800FF00" Canvas.Top="120" Canvas.Left="120"></Ellipse>
    <Ellipse Width="100" Height="100" Fill="#880000FF" Canvas.Top="120" Canvas.Left="200"></Ellipse>
</Canvas>
```

运行结果如图 2-22 所示。

图 2-22　RGB 值示例

2.2.6　画刷类型

大多数可见的 Silverlight 元素都支持一个背景色和一个边框色来填充对象的概念，可以使用背景和边框属性分别声明这两个不同区域的颜色。

1. 边框属性

边框属性值分为边框画刷(Stroke)和边框宽度背景属性(StrokeThickness)。

2. 背景属性

背景属性的属性名称并不完全相同，如 TextBlock、Foreground、Rectangle、Fill、StackPanel、Background 等。

3. 作用范围

在没有边框情况下画刷填充会是被填充对象的全部区域，当为对象添加边框属性并设置边框宽度后，由于边框属性的填充范围是向内的，所以无论设置多大的边框宽度，都不会增大对象本身的宽度，而是缩减对象背景色的可见范围。

4. Brush 类型

Brush 根据作用不同可以分为以下类型。

(1) SolidColorBrush：纯色画刷填充单一颜色到当前区域。

(2) LinearGradientBrush：线性渐变画刷填充线性渐变颜色到当前区域（由一种以上的颜色产生的直线渐变色），并且可以任意搭配两种或两种以上的颜色。

(3) RadialGradientBrush：放射渐变画刷填充放射状渐变颜色到当前区域，类似于 LinearGradient，但是它产生的是一个圆形的渐变颜色，从图形的中心部分向四周逐渐扩展。

(4) ImageBrush：图像画刷使用一幅图像来填充当前区域，可以指定图像的拉伸、缩放或平铺的效果。

(5) VideoBrush：视频画刷使用一个视频来填充当前区域，VideoBrush 采用 Silverlight 中的视频回放特性来实现。

任何元素的着色都是通过专用的色彩填充对象 Brush（画刷）来完成的，画刷包括 SolidColorBrush（用来进行纯色填充）、LinearGradientBrush / RadialGradientBrush（用来进行渐变填充，它们的区别在于前者是线性渐变，后者是放射性渐变）。

5. 纯色画刷

纯色画刷比较简单，就是设置一个颜色值给 Color 属性。但这个颜色值可以设置成：

（1）颜色字符串：如 Red、Blue 等。总共有 256 个命名的颜色串。

（2）RGB 值：如♯0099FF、♯F1F1F1 等。这个与网页中指定的颜色一样。

（3）ARGB 值：如♯880099FF，比 RGB 多了 Alpha 通道信息，用于指定透明度。

设置 RGB 还有一种方式，如♯09F，♯809F，即把每一位重复，这样可得到♯0099FF，♯880099FF。

下面的例子是填充椭圆形：

```
<StackPanel Orientation = "Vertical">
    <!-- Stroke 进行轮廓填充 -->
    <Ellipse Width = "200" Height = "100" Stroke = "Black"
     StrokeThickness = "5">
    <!-- fill 进行背景填充 -->
        <Ellipse.Fill>
            <SolidColorBrush Color = "♯88FF0000"></SolidColorBrush>
        </Ellipse.Fill>
    </Ellipse>
    <Ellipse Width = "200" Height = "100">
        <Ellipse.Fill>
            <SolidColorBrush Color = "♯8800FF00"></SolidColorBrush>
        </Ellipse.Fill>
    </Ellipse>
    <Ellipse Width = "200" Height = "100">
    <Ellipse.Fill>
        <SolidColorBrush Color = "♯880000FF"></SolidColorBrush>
    </Ellipse.Fill>
    </Ellipse>
</StackPanel>
```

运行结果如图 2-23 所示。

图 2-23　纯色画刷填充单一颜色

6. 线性渐变画刷

LinearGradientBrush：线性渐变画刷填充线性渐变颜色到当前区域，可以任意搭配两种或两种以上的颜色。

LinearGradientBrush 中，重要的属性有 GradientStop（倾斜点）、Color（渐变颜色）、Offset（偏移量）、StartPoint（起点坐标）、EndPoint（终点坐标）。

LinearGradientBrush 的渐变色产生一般由多个倾斜点对象组成，其中倾斜点对象的渐变颜色和偏移量这两个属性用来决定渐变色值和它的开始位置。可以简单地把偏移量理解成一个宽度范围，整个偏移量的宽度是 1.0。因此，只要设置好倾斜点的属性，系统就能够自动地在两个倾斜点之间填充好这两种颜色的渐变过程中的颜色。

注：渐变颜色为 red 和 blue，倾斜点的偏移属性表示从 0.0 坐标开始产生渐变到 1.0 坐标结束。例如：

```
<Canvas Width = "400" Height = "320">
    <Rectangle Canvas.Top = "80" Canvas.Left = "50" Width = "300" Height = "150">
        <Rectangle.Fill>
            <LinearGradientBrush>
                <GradientStop Color = "Red" Offset = "0.0"></GradientStop>
                <GradientStop Color = "Blue" Offset = "1.0"></GradientStop>
            </GradientStop>
        </LinearGradientBrush>
    </Rectangle.Fill>
    </Rectangle>
</Canvas>
```

运行结果如图 2-24 所示。

图 2-24　线性渐变画刷填充线性渐变颜色

下面为设置多个倾斜点的效果的例子。

```
<Canvas Width = "400" Height = "320">
    <Rectangle Canvas.Top = "80" Canvas.Left = "50" Width = "300"
    Height = "150">
        <Rectangle.Fill>
            <LinearGradientBrush>
                <GradientStop Color = "Red" Offset = "0.0"></GradientStop>
                <GradientStop Color = "Yellow" Offset = "0.2"></GradientStop>
                <GradientStop Color = "Pink" Offset = "0.4"></GradientStop>
                <GradientStop Color = "Green" Offset = "0.6"></GradientStop>
                <GradientStop Color = "Blue" Offset = "0.8"></GradientStop>
```

```
            <GradientStop Color="Black" Offset="1.0"></GradientStop>
        </LinearGradientBrush>
    </Rectangle.Fill>
  </Rectangle>
</Canvas>
```

运行结果如图 2-25 所示。

图 2-25　设置多个倾斜点的效果

同样,还可以根据起始点和结束点来控制渐变的角度。例如:

```
<Canvas Width="600" Height="420">
    <Rectangle Canvas.Top="10" Canvas.Left="50" Width="300"
       Height="150">
        <Rectangle.Fill>
        <!--水平渐变填充-->
            <LinearGradientBrush StartPoint="0,0.5" EndPoint="1,0.5">
                <GradientStop Color="Green" Offset="0.0"></GradientStop>
                <GradientStop Color="White" Offset="1.0">
                </GradientStop>
            </LinearGradientBrush>
        </Rectangle.Fill>
    </Rectangle>
    <Rectangle Canvas.Top="180" Canvas.Left="50" Width="300"
       Height="150">
        <Rectangle.Fill>
        <!--垂直渐变填充-->
            <LinearGradientBrush StartPoint="0.5,0" EndPoint="0.5,1">
                <GradientStop Color="Green" Offset="0.0">
                </GradientStop>
                <GradientStop Color="White" Offset="1.0">
                </GradientStop>
            </LinearGradientBrush>
        </Rectangle.Fill>
    </Rectangle>
</Canvas>
```

运行效果如图 2-26 所示。

图 2-26　根据起始点和结束点控制渐变的角度

7. 放射渐变画刷

RadialGradientBrush：放射渐变画刷填充放射状渐变颜色到当前区域，类似于 LinearGradientBrush，但是它产生的是一个圆形的渐变颜色，从图形的中间部分向四周组建扩展。

注：不同的倾斜原点和中心值显示不同效果。例如：

```xml
<StackPanel Orientation = "Horizontal" Width = "624" Height = "460">
    <Ellipse Width = "180" Height = "180" Margin = " 50 0 150 0">
        <Ellipse.Fill>
            <RadialGradientBrush GradientOrigin = "0.5,0.5"
                Center = "0.5,0.5">
                <GradientStop Color = "Blue" Offset = "1"></GradientStop>
                <GradientStop Color = "White" Offset = "0"></GradientStop>
            </RadialGradientBrush>
        </Ellipse.Fill>
    </Ellipse>
    <Ellipse Width = "180" Height = "180">
        <Ellipse.Fill>
            <RadialGradientBrush GradientOrigin = "0.3,0.2"
                Center = "0.3,0.2">
                <GradientStop Color = "Blue" Offset = "1"></GradientStop>
                <GradientStop Color = "White" Offset = "0"></GradientStop>
            </RadialGradientBrush>
        </Ellipse.Fill>
    </Ellipse>
</StackPanel>
```

运行效果如图 2-27 所示。

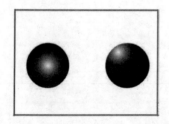

图 2-27　放射渐变画刷填充放射状渐变颜色

2.2.7　使用C#代码绘制画刷

在前面的示例中已经介绍了如何在XAML中使用纯色和渐变画刷元素,本节主要介绍如何在C#中使用它们。

首先介绍一个使用纯色和渐变画刷的应用示例,代码如下:

```
Html 代码
<StackPanel x:Name="LayoutRoot" Background="White" Orientation="Vertical">
        <Ellipse x:Name="ellipse" Width="260" Height="260" Fill="GreenYellow"/>
        <TextBlock x:Name="txb_txb" FontSize="30" Text="使用C#绘制Brush"/>
    </StackPanel>
C#:
Csharp 代码
public partial class DrawBruseWithCSharp : UserControl
{
    public DrawBruseWithCSharp()
    {
        InitializeComponent();

        //绘制单色填充
        ellipse.Stroke = new SolidColorBrush(Colors.Black);
        ellipse.StrokeThickness = 3;
        //绘制渐变填充
        LinearGradientBrush lgb = new LinearGradientBrush();
        lgb.GradientStops.Add(new GradientStop() { Color = Colors.Green, Offset = 0});
        lgb.GradientStops.Add(new GradientStop() { Color = Colors.Yellow,Offset = 1});
        txb_txb.Foreground = lgb;
        ellipse.Fill = lgb;
    }
}
```

运行结果如图2-28所示。

图2-28　使用C#代码绘制画刷

总　结

本章介绍了由Silverlight所提供的图像API,以及派生自Shape的对象(Line、Rectangle、Ellipse、Polygon、Polyline和Path)。为了将Path对象渲染到屏幕上,Path对象需要Geometry对象,并且Silverlight直接提供了多个派生自Geometry的对象,如RectangleGeometry、EllipseGeometry、LineGeometry和PathGeometry。这些对象还可以组合在一个GeometryGroup对象中,以创建更加复杂的形状。

本章还介绍了 Brush 对象,以及不同类型的 Brush 对象,这些对象分别使得既可以用某种纯单一的颜色(SolidColorBrush)绘制图像,也可以使用变化的不同颜色绘制图像,既可以沿直线渐变的颜色(LinearGradientBrush),还可以从某个中心点向外扩散的颜色渐变(RadialGradientBrush)。

通过学习线性绘图和几何绘图,可以对 Silverlight 中的绘图对象有了一个全面的认识和理解。根据需求选择适合的绘图元素可以使 Silverlight 界面更加美观。

作 业

1. 简述路径标记语法中的指令,并说出有几个绘制指令。
2. 在上面列举的指令中筛选一个指令,绘制一个嘴巴的图形,效果如图 2-29 所示。

图 2-29　用 Path 绘制笑脸

第3章

图像与视觉特效

学习目标

➢ 掌握 Silverlight 中的基本图形特效
➢ 掌握在 C# 中控件图形的特效
➢ 灵活使用 3D Effects 特效

图像元素和图像画刷元素的作用都是将一幅图像文件在 Silverlight 中显示出来,两者作用相同但呈现形式不同,其中图像对象是单纯用来显示位图文件,而 ImageBrush 可以作为任何支持 Brush 填充元素的填充内容,所以将它们放在同一章。本章还对 Silverlight 3 的 3D 特性进行了详细的分析。

3.1 图像对象

图像(Image)对象是 Silverlight 中显示图片的主要元素,其作用是显示一幅位图文件,下面介绍一个简单的 Image 应用例子,使用一个 Image 元素来显示一幅游戏的图片,图片格式为 PNG,示例代码如下:

```
XAML:
< Grid x:Name = "LayoutRoot" Background = "AliceBlue">
        < Image Source = "Images/leaf1.jpg"></Image>
</Grid>
```

运行结果如图 3-1 所示。

这段代码很简单,Grid 中包含一个 Image 元素,并声明了它的 Source 属性,Image 元素在默认情况下会完整呈现位图的大小,可以指定它的 Width 和 Height 属性来修改图片的大小,将示例代码修改后如下:

```
XAML:
    <Grid x:Name = "LayoutRoot" Background = "AliceBlue">
        <Image Source = "../Images/Halo640.JPG" Width = "500"
        Height = "300"></Image>
</Grid>
```

示例的运行结果如图 3-2 所示。

图 3-1　使用图像(Image)对象

图 3-2　通过修改 Width 和 Height 改变图片大小

提 示

当对 Image 元素只设置了 Width 或 Height 中的某一个属性值，并且没有定义 Stretch 属性时，Image 将根据 Width 或 Height 之一的尺寸来等比缩放图像。

3.1.1　图片拉伸属性

当源位图的实际大小与 Image 元素声明的 width 和 height 属性的尺寸不同时，可以使用 Stretch(拉伸)属性控制图像的拉伸方式，拉伸属性基于图片的 Width 和 Height 属性，这里使用微软 Silverlight 的 Logo 作为源图，示例代码如下：

```
XAML:
<StackPanel Background = "White">
    <!-- 无拉伸 -->
    <Image Source = "../Images/sllogo.jpg"
        Width = "85" Height = "85" Margin = "5"
        Stretch = "None"/>
    <!-- 充拉伸 -->
    <Image Source = "../Images/sllogo.jpg"
        Width == "85" Height = "85" Margin = "5" Stretch"Fill"/>
    <!-- 等比拉伸 -->
    <Image Source = "../Images/sllogo.jpg"
        Width = "85" Height = "85" Margin = "5"
Stretch = "Uniform"/>
    <!-- 等比并填拉伸 -->
    <Image Source = "../Images/sllogo.jpg"
        Width = "85" Height = "85" Margin = "5"
```

```
            Stretch = "UniformToFill"/>
    </StackPanel>
```

运行结果如图 3-3 所示。

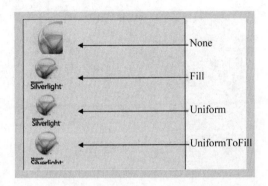

图 3-3 使用 Stretch(拉伸)属性控制图像的拉伸方式

Image 元素的 Stretch 属性值类型是枚举型,属性值分别为原始尺寸(None)、填充拉伸(Fill)、等比拉伸(Uniform)和等比拉伸填充(UniformToFill)。

从图中可以发现,当设置 Image 的 Stretch 属性为 None 和 UniformToFill 值时,Logo 图片没有被完整显示,这是因为图片拉伸后的大小超出了 Width 和 Height 属性指定的范围,所以超出的部分被忽略了。

3.1.2 图像画刷

在 Silverlight 中,除了可以直接使用 Image 元素来显示图像以外,还可以使用一幅图像来填充一个 Silverlight 元素,只要这个元素的属性可以接受 Brush 类型,这就是图像画刷(ImageBrush)。ImageBrush 和 Image 一样可以指定图像的拉伸效果,下面看一个图像画刷实例,是将一幅图片填充到一个矩形元素中作为背景图片,示例代码如下:

```
XAML:
< UserControl x:Class = "Sample.imagebrush"
xmlns = "http://schemas.microsoft.com/winfx/2006/xaml/presentation"
xmlns:x = "http://schemas.microsoft.com/winfx/2006/xaml"
Width = "640" Height = "340">
    < Grid x:Name = "LayoutRoot" Background = "White">
        < Rectangle Width = "600" Height = "300"
RadiusX = "50" RadiusY = "50">
            < Rectangle.Fill >
                < ImageBrush ImageSource = "Images/Halo640.PNG"/>
            </Rectangle.Fill >
        </Rectangle >
```

```
            </Grid>
    </UserControl>
```

运行结果如图 3-4 所示。

图 3-4　使用图像画刷填充矩形

由于这里的 Rectangle.Fill 属性填充的不是单一的颜色,所以<ImageBrush>元素被放在<Rectangle.Fill>…</Rectangle.Fill>之间,矩形元素的 RadiusX 和 RadiusY 属性用来声明矩形的圆角边缘,通过改变 RadiusX 和 RadiusY 属性值可以调整圆角的程度。

ImageBrush 和 Image 一样支持 Stretch 属性,当声明拉伸属性值为 None,并且图像的尺寸小于被填充元素的尺寸时,可以使用 AlignmentX 和 AlignmentY 属性来改变图像填充的位置。

3.2　在 C#中使用图像

与 XAML 不同,在 C#中使用 Image 对象不可以直接对 Image.Source 属性赋值,需要先创建 URI 和 BitmapImage 对象,示例如下:

```
C#:
namespace Sample
{
    public partial class imagesharp : UserControl
    {
        public imagesharp()
        {
            InitializeComponent();
            //声明图片路径
            Uri uri = new Uri("Images/Halo640.JPG");
            //定义图片源
            BitmapImage bitmap = new BitmapImage(uri);
```

```
            //为 Source 属性赋值
            Image.Source = bitmap;
        }
    }
}
```

其中 Uri 对象为 BitmapImage 提供图片路径，而 BitmapImage 为 Image.Source 提供图片源，它是为 Image.Source 和 ImageBrush.ImageSource 属性提供图像源的专用对象。

3.3 使用 BitmapImage 的下载事件

在使用 Image 对象的过程中，对于在一个 Silverlight 应用程序中同时显示大量图片，或者图片体积比较大时，在没有下载完整个图片之前客户端是看不到它的，这时可以通过图片的下载进度事件来获取图片的已下载大小，并告知浏览者，示例如下：

```
XAML：
<UserControl x:Class="Sample.imagedown"
xmlns="http://schemas.microsoft.com/winfx/2006/xaml/presentation"
xmlns:x="http://schemas.microsoft.com/winfx/2006/xaml"
Width="400" Height="300">
    <Grid x:Name="LayoutRoot" Background="White">
        <Image x:Name="image"/>
        <ProgressBar x:Name="progressBar"
Width="270" Height="40"/>
    </Grid>
</UserControl>
C#：

namespace Sample
{
    public partial class imagedown : UserControl
    {
        public imagedown()
        {
            InitializeComponent();
            progressBar.Value = 0;
            //声明图片路径
            Uri uri = new Uri("Images/myLogo.jpg",
            UriKind.RelativeOrAbsolute);
            //创建位图对象
            BitmapImage bitmap = new BitmapImage();
            //定义图片下载事件
            bitmap.DownloadProgress +=
            new EventHandler<DownloadProgressEventArgs>
            (bitmap_DownloadProgress);
```

```
            //定义图片源
            bitmap.UriSource = uri;
            //为Source属性赋值
            image.Source = bitmap;
        }
        //图片源下载进度事件处理
        void bitmap_DownloadProgress(object sender,
          DownloadProgressEventArgs e)
        {
            //将下载进度反映到进度条控件
            progressBar.Value = (double)e.Progress;
        }
    }
}
```

在XAML界面使用了一个进度条控件和一个Image对象,通过BitmapImage(位图对象)的DownloadProgress事件来获取图片的下载进度,并将结果传给进度条控件的Value值,使用BitmapImage对象需要先引用System.Windows.Media.Imaging命名空间。

3.3.1 使用WriteableBitmap绘制位图

Silverlight 3引入WriteableBitmap类,它提供了一个可写的位图类,允许使用像素单位生成一个位图,由于WriteableBitmap位于System.Windows.Media.Imaging命名空间,并派生于BitmapSource类,所以它属于图像的范畴。

WriteableBitmap和BitmapImage类似,它们都可以为Image对象提供Source属性值。WriteableBitmap示例代码如下:

```
C#:
//创建一个Image对象
Private Image myBitmap = new Image();
Private cons tint imageWidth = 200;
Private cons tint imageHeight = 200;
//构造函数
Public BuildBitmapSample()
{
    InitializeComponent();
    myBitmap.Width = imageWidth;
    myBitmap.Height = imageHeight;
    LayoutRoot.Children.Add(myBitmap);
    BuildBitmap();
}
//创建图形
Private void BuildBitmap()
{
    //创建可写位图对象
    WriteableBitmap b = new WriteableBitmap(imageWidth,imageHeight);
    for(int x = 0;x < imageWidth;x++)
```

```
        {
            Byte[ ]rgb = new byte[4];
            Rgb[0] = (byte)(x % 255);
            Rgb[1] = (byte)(y % 255);
            Rgb[2] = (byte)(x * y % 255);
            Rgb[3] = 0;
            Int pixelValue = BitConverter.ToInt32(rgb,0);
            b.Pixels[y * imageWidth + x] = pixelValue;
        }
    }
    myBitmap.Source = b;
}
```

在这个示例的 XAML 中，除了一个基本 Grid 之外没有其他的 XAML 元素，WriteableBitmap 和 BitmapImage 里没有 XAML，在后台代码中引用两个条件循环语句,通过程序设置 Pbgra32 格式的颜色,通过 rgb 设置了一个数组,分别执行蓝色、绿色、红色的计算。

提 示

使用 WriteableBitmap 生成位图需要具备一定的图形像素理论,这里只是展现了 WriteableBitmap 的效果,后面还会介绍 WriteableBitmap 的 Render 功能。

3.3.2 文本画刷应用

画刷的用途很广,除了可以用来填充矩形、椭圆形区域之外,还可作用于其他任何支持 Brush 对象的属性值上,如文本元素 TextBlock 的 Foreground 属性。下面介绍一个使用 Bursh 对象填充文本对象的示例,代码如下：

```
XAML:
< Canvas Background = "AliceBlue">
        < TextBlock Text = "图像画刷">
                    FoutFamily = "Verdana"
                    FontSize = "90"
                    FontWeight = "Bold"
                    FontStyle = "Italic">
        < TextBlock.Foreground >
                    <! - 使用图像画刷填充 Foreground 属性 - >
                    < ImageBrush
                        imageSource = "../Images/SilverlightBlack.jpg"/>
        </TextBlock.Foreground >
        </TextBlock >
</Canvas>
```

运行结果如图 3-5 所示。

图 3-8 使用透明遮罩属性

Color 值决定,当 Color 值为♯00000000 时对象完全透明,相反,当 Color 值为♯FF000000 时对象不透明。

下面使用线性渐变画刷实现透明遮罩效果,示例代码如下:

```
XAML:
<UserControl x:Class = "Sample.OpacityMaskB"
xmlns = "http://schemas.microsoft.com/winfx/2006/xaml/presentation"
xmlns:x = "http://schemas.microsoft.com/winfx/2006/xaml"
Width = "640" Height = "480">
    <Canvas Background = "White">
        <Image Source = "Images/Halo640.PNG">
            <Image.OpacityMask>
                <!-- 使用线性渐变画刷 -->
                <LinearGradientBrush>
                    <GradientStop Offset = "0" Color = "♯00000000" />
                    <GradientStop Offset = "1" Color = "♯FF000000" />
                </LinearGradientBrush>
            </Image.OpacityMask>
        </Image>
    </Canvas>
</UserControl>
```

运行结果如图 3-9 所示。

图 3-9 渐变颜色(透明度)

3.6 裁剪特效

裁剪特效(Clip)属性允许用户将任何支持裁剪属性的 Silverlight 元素进行剪切,剪切的形状由 Clip 属性内包含的 Geometry 形状决定,用户可以使用几何矩形、几何椭圆形和几何路径等。关于几何绘图,前面已经介绍过,使用 Clip 属性仅需要声明 Clip 属性的几何形状即可,示例代码如下:

```xaml
XAML:
Width = "600" Height = "600">
    <Grid Background = "White">
        <Grid.RowDefinitions>
            <RowDefinition Height = "300"/>
            <RowDefinition Height = "300"/>
        </Grid.RowDefinitions>
        <Grid.ColumnDefinitions>
            <ColumnDefinition/>
        </Grid.ColumnDefinitions>
        <!-- 原始 Image 位图 -->
        <Image Width = "320" Height = "240"
        Grid.Row = "0" Grid.Column = "0"
        Source = "Images/Halo640.PNG"/>
        <!-- 声明 Clip 属性的 Image 位图 -->
        <Image Width = "320" Height = "240"
        Grid.Row = "1" Grid.Column = "0"
        Source = "Images/Halo640.PNG">

            <Image.Clip>
                <!-- Clip 的范围是几何椭圆形 -->
                <EllipseGeometry Center = "200,120"
                RadiusX = "100" RadiusY = "120">
                </EllipseGeometry>
            </Image.Clip>

        </Image>
    </Grid>
</UserControl>
```

运行结果如图 3-10 所示。

上面的示例声明了两个 Image 元素,分别是使用 Clip 剪切的图片和原始图像,很明显,原始图像是矩形的,而使用 Clip 属性后的图像变成椭圆形,这个椭圆形是由 EllipseGeometry 产生的,椭圆形以外的边缘被剪切掉了。

3.7.2 RotateTransform 对象

旋转变形时,RenderTransform 属性用来产生旋转效果的元素,示例代码如下:

```
XAML:
<Canvas x:Name="LayoutRoot" Background="Black">
    <!--声明原 Image-->
    <Image Source="../images/jiqiren.jpg" Canvas.Left="335" Canvas.Top="80"></Image>
    <!--new Image-->
    <Image Source="../images/jiqiren.jpg" Canvas.Left="335" Canvas.Top="80">
        <!--声明 RenderTransform 属性-->
        <Image.RenderTransform>
            <!--使用 RotateTransform 旋转对象-->
            <RotateTransform CenterX="25" CenterY="25" Angle="45" />
        </Image.RenderTransform>
    </Image>
    <Image Source="../images/jiqiren.jpg" Canvas.Left="335" Canvas.Top="80">
        <!--声明 RenderTransform 属性-->
        <Image.RenderTransform>
            <!--使用 RotateTransform 旋转对象-->
            <RotateTransform CenterX="25" CenterY="25" Angle="90" />
        </Image.RenderTransform>
    </Image>
</Canvas>
```

运行结果如图 3-12 所示。

图 3-12 使用 RotateTransform 对象进行旋转

上面例子中的 Angle(角度)属性用来指定旋转的角度值,声明 Angle 属性为 45 时代表将图像旋转 45°角,示例中分别对原始图像旋转 45°和 90°。旋转变形元素的 CenterX 和 CenterY 属性用来声明旋转的中心点坐标,对象根据这个中心点坐标进行旋转,旋转变形常用在各种旋转动画效果中。

3.7.3　ScaleTransform 对象

当不希望通过改变某个对象的 Width 和 Height 值的方式来改变其尺寸时，可以考虑使用缩放变形素对它进行不同比例的缩放，示例代码如下：

```
XAML:
< Canvas x:Name = "LayoutRoot" Background = "Black">
        <! -- 声明原 Image -- >
        < Image Source = "../images/jiqiren.jpg" Canvas.Left = "335" Opacity = "0.5" Canvas.
Top = "80"></Image >
        < Image Source = "../images/jiqiren.jpg" Canvas.Left = "335" Canvas.Top = "80">
            < Image.RenderTransform >
                <! -- ScaleTransform 更改对象的大小 -- >
                < ScaleTransform CenterX = "0" CenterY = "0" ScaleX = "0.7" ScaleY = "0.7" />
            </Image.RenderTransform >
        </ Image >
</Canvas >
```

运行结果如图 3-13 所示。

图 3-13　使用 ScaleTransform 对象进行缩放

首先原始图像 Opacity 属性声明为半透明效果，缩放变形元素的 ScaleX 和 ScaleY 属性分别指定产生缩放的横纵的百分比，这个比例是相对于被缩放对象的原始尺寸而言的，示例中的 ScaleX 和 ScaleY 属性值均为 0.5，表示相对于原始大小缩小一半。当属性值为 1.5 时，对象的尺寸将放大为 1.5 倍，当属性值为 1 时，对象的尺寸不会发生任何变化。

缩放变形元素同样也包含 CenterX 和 CenterY 属性，用来指定产生缩放效果的中心点坐标。缩放变形元素常用于缩放动画效果中。

3.7.4　SkewTransform 对象

扭曲变形的作用是使任意一个 Silverlight 元素产生扭曲变化，可以使其产生水平和垂直的倾斜效果，或水平和垂直扭曲效果加在一起，使用扭曲变形元素仅需要声明它的 AngleX 和 AngleY 属性值即可，示例代码如下：

M11 Default: 1.0	M12 Default: 0.0	0.0
M21 Default: 0.0	M22 Default: 1.0	0.0
OffsetX Default: 0.0	OffsetY Default: 0.0	1.0

图 3-16　矩形变形对象的组成原理

该矩阵的最后一列值是固定的，不会改变。如果修改矩阵中的 OffsetX 值，元素将会在 X 轴上进行移动；如果修改 OffsetY 值，元素将在 Y 轴上移动；如果修改 M22 为 2，元素的高度将会拉伸到 2 倍。通过修改该矩阵，能实现前面提到的几种简单变换的所有功能，事实上前几种简单变换只是矩阵变换的特例而已，单独使用 MatrixTransform 对象，可以实现所有的变换，示例代码如下：

```xaml
XAML
<Canvas x:Name="LayoutRoot" Background="White">
    <TextBlock Text="欢迎进入 Silverlight 世界"
    Canvas.Top="30" Canvas.Left="30"
    Opacity="0.5">
</TextBlock>
    <TextBlock Text="欢迎进入 Silverlight 世界"
    Canvas.Top="30" Canvas.Left="30"
    Foreground="OrangeRed">
    <TextBlock.RenderTransform>
    <MatrixTransform>
    <MatrixTransform.Matrix>
    <Matrix M11="1" M12="0.3"
    M21="0.3" M22="0.8"
    OffsetX="10" OffsetY="20"/>
    </MatrixTransform.Matrix>
    </MatrixTransform>
    </TextBlock.RenderTransform>
    </TextBlock>
</Canvas>
```

运行效果如图 3-17 所示。

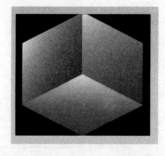

图 3-17　使用 MatrixTransform 变形元素模拟一个 3D 立体盒子效果

上面示例中的立体盒子实际是由 3 个矩形变形后组成的,盒子的每一面都由一个矩形变形后产生,由于 MatrixTransform 属性值产生了不同的矩形变形效果,所以把它们组合起来有一种 3D 效果。

3.7.7 在 C♯中应用变形对象

在 C♯代码中,Silverlight 允许对所有支持 RenderTransform 属性的对象应用变形对象,可以在 C♯中创建一种或多种变形对象,变形对象不同的属性值决定它产生不同的效果。

其实这些变形对象的原理都是使用数学矩阵理论来产生的,在 Silverlight 中无须了解这些数学上复杂的理论原理,只需要创建并应用相应的变形对象即可。下面通过一个按钮来实现对图形的旋转变形,示例代码如下:

```
XAML:
< Grid x:Name = "LayoutRoot" Background = "White">
    <! -- 图片对象 -->
    < Image X:Name = "image" Stretch = "None" Source = "../Images/MP.jpg">
    <! -- 按钮实现旋转 -->
    < Button Width = "100" Height = "40" Content = "0"
            Click = "Button_Click" FontSize = "18"/>
</Grid>
C♯:
//声明角度变量
Private double angle = 0;
//声明变形对象
Private Rotate Transform rotate = new RotateTransform();
Public transSharpA()
{
        InitializeComponent();
}
Private void Button_Click(object sender,RoutedEventArgs e)
{
        //增加旋转角度
        Angle += 45;
        //引用按钮对象
        Button button = sender as Button;
        //设置按钮文本
        Button.Content = angle.ToString();
        //应用旋转角度值
        Rotate.Angle = angle;
        //设置旋转中心点
        Rotate.CenterX = 100;
        Rotate.CenterY = 100;
        //应用变形对象到 image 对象
        Image.RenderTransform = rotate;
}
```

运行结果如图 3-18～图 3-20 所示。

```
                <LinearGradientBrush>
                    <GradientStop Offset = "0" Color = "White" />
                    <GradientStop Offset = "1" Color = "Black" />
                </LinearGradientBrush>
            </StackPanel.Background>
            <!-- 内容 -->
            <TextBlock Text = "Silverlight 3D Effects" FontSize = "20" />
            <TextBlock Text = "UserName:" Margin = "10" />
            <TextBox Margin = "10" Width = "200" />
            <TextBlock Text = "PassWord:" Margin = "10" />
            <TextBox Margin = "10" Width = "200" />
            <Button Width = "100" Height = "30" Content = "Login" />
        </StackPanel>
    </Grid>
```

运行结果如图 3-22 所示。

Projection 除了对 X、Y、Z 轴的值旋转之外，还允许调整 Projection 对 Silverlight 目标对象进行基于三维空间的定位。

（1）LocalOffsetX 沿旋转对象的 X 轴定位一个对象。

（2）LocalOffsetY 沿旋转对象的 Y 轴定位一个对象。

（3）LocalOffsetZ 沿旋转对象的 Z 轴定位一个对象。

（4）GlobalOffsetX 沿屏幕对齐的 X 轴定位一个对象。

（5）GlobalOffsetY 沿屏幕对齐的 Y 轴定位一个对象。

图 3-22　使用 3D 特效的效果

（6）GlobalOffsetZ 沿屏幕对齐的 Z 轴定位一个对象。

为了更好地体会对象的 X 轴、Y 轴、Z 轴的旋转效果，下面设计一个例子，通过拖动进度条来更改 X 轴、Y 轴、Z 轴的值。代码如下：

```
XAML:
<Grid x:Name = "LayoutRoot">
    <Grid.RowDefinitions>
        <RowDefinition Height = "*" />
        <RowDefinition Height = "Auto" />
        <RowDefinition Height = "Auto" />
        <RowDefinition Height = "Auto" />
    </Grid.RowDefinitions>
    <Grid.ColumnDefinitions>
        <ColumnDefinition Width = "*" />
        <ColumnDefinition Width = "Auto" />
    </Grid.ColumnDefinitions>
    <Grid.Background>
        <ImageBrush ImageSource = "../images/widows7.jpg" />
    </Grid.Background>
    <!-- 3D 元素 -->
    <Grid Width = "200" Height = "160" Background = "Green" >
```

```xml
<TextBlock Foreground = "White" Text = "Windows
 Phone 7" Grid.Column = "0" FontFamily = "Arial" FontSize
  = "22" Margin = "6,58,6,20">
</TextBlock>
<Grid.Effect>
    <DropShadowEffect Direction = "50" BlurRadius
     = "60" Color = "Green" Opacity = "0.8" />
</Grid.Effect>
<Grid.Projection>
    <PlaneProjection x:Name = "prots" />
</Grid.Projection>
</Grid>
<!-- Slider 控件和元素之间的 TwoWay 数据绑定用来设置 Projection 属性 -->
<Slider Height = "23" Grid.Column = "0" Grid.Row = "1" Maximum = "180"
 Minimum = "-180" Value = "{Binding RotationX,Mode
= TwoWay,ElementName = prots}" />
<Slider Height = "23" Grid.Column = "0" Grid.Row = "2" Maximum
 = "180" Minimum = "-180" Value = "{Binding RotationY,Mode
= TwoWay,ElementName = prots}"/>
<Slider Height = "23" Grid.Column = "0" Grid.Row = "3" Maximum = "180"
 Minimum = "-180" Value = "{Binding RotationZ,Mode =
TwoWay,ElementName = prots}"/>
<!-- 用 TextBox 显示坐标 -->
<TextBox Height = "23" Width = "150" Grid.Column = "1" Grid.Row = "1"
 Margin = "10" IsReadOnly = "True" Text = "{Binding RotationX,Mode
= TwoWay,ElementName = prots}"/>
<TextBox Height = "23" Width = "150" Grid.Column = "1" Grid.Row = "2"
 Margin = "10" IsReadOnly = "True" Text = "{Binding RotationY,Mode
= TwoWay,ElementName = prots}"/>
<TextBox Height = "23" Width = "150" Grid.Column = "1" Grid.Row = "3"
 Margin = "10" IsReadOnly = "True" Text = "{Binding RotationZ,Mode =
TwoWay,ElementName = prots}"/>
</Grid>
<Image.Projection>
            <PlaneProjection RotationX = "-60" />
        </Image.Projection>
    </Image>
</Grid>
```

运行结果如果 3-23 所示。

本示例设置 Slider 控件的最大值为 180、最小值为-180，通过 Slider 控件可以在程序运行时任意单独、组合调整 PlaneProjection 元素不同的坐标属性值，可根据以上实例添加更多的 Slider 控件来对 PlaneProjection 对象的其他属性进行控制，从而进一步了解它们的作用。

与此同时，在 Expression Blend 最新的 3.0 版本中新加入了 Silverlight 界面元素的三维属性控制面板，通过 Blend 3 可以对界面中任何元素的 PlaneProjection 属性值进行调整，设置元素的三维视角。

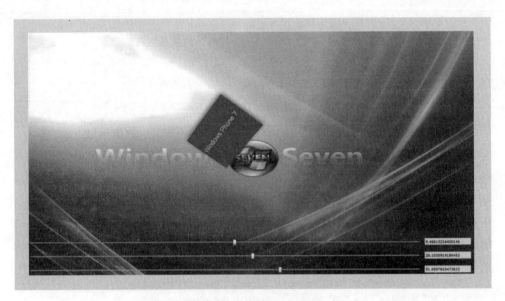

图 3-23　示例的运行结果

3.7.9　关于 Element-To-Element Binding

前面的 3D Effects 示例中使用了 Silverlight 3 新添加的 Element-To-Element Binding 特性，所以这里继续略做说明。Element-To-Element Binding 起初是 WPF 所支持的数据绑定方式，在 Silverlight 3 中也被包含了进来，通过 Element-To-Element Binding 可以实现不同的元素属性值之间的数据绑定，也就是 UI 和 UI 之间属性的数据实时交换。

Element-To-Element Binding 格式：

```
{Binding 属性名,Mode＝.绑定模式,ElementName＝绑定元素}
```

Element-To-Element Binding 和一般的 Silverlight 绑定一样也支持 OneWay 和 TwoWay 的绑定方向。

3.8　Silverlight 3 Effect 特效

Silverlight 3 新增了 System.Windows.Media.Effects 命名空间，在它下面包含一个 Effect 类，是 Silverlight 3 中专门用来产生各种特效的类。本节主要介绍 Effect 类中的两个重要成员：BlurEffect 和 DropShadowEffect。它们是一组非常实用的特效，下面分别介绍其作用。

3.8.1　BlurEffect

Silverlight 中的每个对象都支持添加模糊和阴影效果，在 Blend 工具中通过"外观"面板可以直接可视化地进行设计完成模糊和阴影效果的添加，以及效果参数的调整。

使用相册开发中的一个典型实例来分析,例如照片默认缩小为一定的比例并添加了模糊效果呈现在相片列表中,当鼠标指向照片时将照片进行放大显示(放大图片可通过缩放动画实现,具体可查阅《Silverlight & Blend 动画设计系列三:缩放动画(ScaleTransform)》),并动态改变其模糊效果值为0以清晰地查看照片效果。定义图片的代码如下:

```
< Image Height = "240" x:Name = "Flower" Width = "320" Source
 = "yellowFlower.jpg" Opacity = "1"
    Canvas.Left = "240" Canvas.Top = "180" RenderTransformOrigin
    = "0.5,0.5"
    MouseEnter = "Flower_MouseEnter" MouseLeave = "Flower_MouseLeave"
     OpacityMask = "{x:Null}">
    < Image.RenderTransform >
        < TransformGroup >
            < ScaleTransform/>
            < SkewTransform/>
            < RotateTransform/>
            < TranslateTransform/>
        </TransformGroup >
    </Image.RenderTransform >
    < Image.Effect >
        < BlurEffect/><!-- 为该图片添加了模糊效果 -->
    </Image.Effect >
</Image >
```

可以通过两个动画来处理,一个动画将图片按缩放比例进行放大(ScaleX,ScaleY:1-->2)同时改变其模糊值为0(Radius:5-->0),另一个动画则相反,代码如下:

```
< Storyboard x:Name = "Flower_Enter">
    < DoubleAnimation BeginTime = "00:00:00" Storyboard.TargetName
    = "Flower" Duration = "00:00:00.50" To = "2"

Storyboard.TargetProperty = "(UIElement.RenderTransform).(Transform Group.Children)[0].
(ScaleTransform.ScaleX)">
    </DoubleAnimation >
    < DoubleAnimation BeginTime = "00:00:00" Storyboard.TargetName
    = "Flower" Duration = "00:00:00.50" To = "2"

Storyboard.TargetProperty = "(UIElement.RenderTransform).(TransformGroup.Children)[0].
(ScaleTransform.ScaleY)">
    </DoubleAnimation >
    < DoubleAnimation BeginTime = "00:00:00" Storyboard.TargetName
    = "Flower" Duration = "00:00:00.50" To = "0"

Storyboard.TargetProperty = "(UIElement.Effect).(BlurEffect.Radius)">
    </DoubleAnimation >
</Storyboard >
< Storyboard x:Name = "Flower_Level">
    < DoubleAnimation BeginTime = "00:00:00" Storyboard.TargetName
```

```
           = "Flower" Duration = "00:00:00.50" To = "1"

Storyboard.TargetProperty = "(UIElement.RenderTransform).(TransformGroup.Children)[0].
(ScaleTransform.ScaleX)">
    </DoubleAnimation>
    <DoubleAnimation BeginTime = "00:00:00" Storyboard.TargetName
     = "Flower" Duration = "00:00:00.50" To = "1"

Storyboard.TargetProperty = "(UIElement.RenderTransform).(TransformGroup.Children)[0].
(ScaleTransform.ScaleY)">
    </DoubleAnimation>
    <DoubleAnimation BeginTime = "00:00:00" Storyboard.TargetName
     = "Flower" Duration = "00:00:00.5000000" To = "5"

Storyboard.TargetProperty = "(UIElement.Effect).(BlurEffect.Radius)">
    </DoubleAnimation>
</Storyboard>
```

通过鼠标事件（MouseEnter,MouseLeave）动态地触发上面定义的两个动画的执行即可达到预期的目的，代码如下：

```
private void Flower_MouseEnter(object sender, System.Windows.Input.MouseEventArgs e)
{
    //TODO: Add event handler implementation here.
    this.Flower_Enter.Begin();
}

private void Flower_MouseLeave(object sender, System.Windows.Input.MouseEventArgs e)
{
    //TODO: Add event handler implementation here.
    this.Flower_Level.Begin();
}
```

运行结果如图 3-24 所示。

图 3-24　运行结果

3.8.2 DropShadowEffect

在 Silverlight 中应用阴影效果和模糊效果一样简单，可使用添加模糊效果的方法对元素添加阴影效果，需要注意的是设置阴影效果的相关属性。

（1）BlurRadius：模糊半径。
（2）Color：填充颜色。
（3）Direction：方向。
（4）Opacity：透明度。
（5）ShadowDepth：阴影深度。

通过以上几个属性选项的设置即可完成阴影效果的设计。运行结果如图 3-25 所示。

图 3-25　阴影效果（DropShadowEffect）

对应生成的 XAML 代码如下：

```
< Image Height = "240" x:Name = "Flower" Width = "320" Canvas.Left = "240" Canvas.Top = "180"
    Source = "yellowFlower.jpg" Stretch = "Fill" Cursor = "Hand">
    < Image.Effect >
        < DropShadowEffect x:Name = "FlowerShadow"
        BlurRadius = "18"
        ShadowDepth = "27"
        Opacity = "0.6"
        Direction = "321"/>
    </Image.Effect>
</Image>
```

总　结

本章主要介绍了 Silverlight 中的图像和图像画刷的应用；围绕图像和图像画刷介绍了 Silverlight 中的特效属性，主要包括透明、透明遮罩、剪切和变形元素，这些特效属性既可以单独使用，又可以多种效果组合使用。

在本章的后半部分重点介绍了 Silverlight 3 中的新增特效，其中包括 3D Effects 投影特效和模糊特效。

 作 业

1. 在 RenderTransform 中包含哪些成员？每个成员的作用是什么？
2. 编写代码，实现模拟 Windows 窗体的效果，如图 3-26 所示。要求如下：
(1) 实现矩形框的拖曳。
(2) 实现矩形框的放大和缩小。

图 3-26　单击按钮放大图片

第4章 动画与多媒体

> **学习目标**
> - 了解触发器
> - 掌握3种线性插值动画
> - 掌握3种关键帧动画
> - 掌握 MediaElement 的使用

4.1 故事板和事件触发器

1. Storyboard

Storyboard(故事板)是 Silverlight 动画的基本单元,用来分配动画时间,可以使用同一个故事板对象产生一种或多种动画效果,并且允许控制动画的播放、暂停、停止以及在何时何地播放。

使用故事板时,必须指定 TargetProperty(目标)属性和 TargetName(目标名称)属性,这两个属性把故事板和所有产生的动画效果衔接起来,起到了桥梁的作用。下面是一个完整的 Storyboard 动画代码。

```
<Storyboard>
    <DoubleAnimation Storyboard.TargetName = "ellipse1"
        Storyboard.TargetProperty = "Width"
        From = "150" To = "300"
        Duration = "0:0:3" />
</Storyboard>
```

上面示例中的动画效果是产生一个变形的椭圆,代码声明了一个故事板和一个 DoubleAnimation 类型的动画对象,DoubleAnimation 动画元素指定了 TargetName(作用目标)

和作用属性(TargetProperty),其中作用目标的值为 Ellipse1(声明了 DoubleAnimation 动画效果的作用目标为 ellipse1),TargetProperty 值为 Width(声明 DoubleAnimation 的动画作用的属性为椭圆形的宽度值)。

2. EventTrigger

当完成一个故事板定义并声明了动画类型之后,这个动画并不能在 XAML 页面加载后自动播放,因为并没有指定动画播放的开始事件,此时需要使用 EventTrigger(事件触发器)对象,然后通过事件触发器播放 BeginStoryboard 故事板的动画效果,接下来将前面的示例加入事件触发元素中,组成一个完整的播放动画。

```
XAML 代码:
<UserControl

xmlns = "http://schemas.microsoft.com/winfx/2006/xaml/presentation"
    xmlns:x = "http://schemas.microsoft.com/winfx/2006/xaml"
    x:Class = "Anim.MainPage"
    Width = "640" Height = "480">

    <Canvas Background = "AliceBlue">
        <Ellipse x:Name = "ellipse1" Fill = "GreenYellow"
                Width = "150" Height = "200" />
        <Canvas.Triggers>
            <!-- 创建触发器 -->
            <EventTrigger RoutedEvent = "Canvas.Loaded">
                <EventTrigger.Actions>
                    <BeginStoryboard>
                        <Storyboard>
                            <DoubleAnimation
                                Storyboard.TargetName = "ellipse1"
                                Storyboard.TargetProperty = "Width"
                                From = "150" To = "300" Duration = "0:0:3" />
                        </Storyboard>
                    </BeginStoryboard>
                </EventTrigger.Actions>
            </EventTrigger>
        </Canvas.Triggers>
    </Canvas>
</UserControl>
```

上面的代码声明了一个 Canvas.Triggers(触发器),然后声明 RoutedEvent(事件通道)为 Canvas.Loaded(Canvas 加载完成后触发的事件)。EventTrigger.Actions(事件行为)包含了 BeginStoryboard,用来开始播放故事板里的动画对象。简而言之,就是当这个 XAML 的画布加载完成就立刻播放故事板包含的动画效果。

除了在 XAML 代码中使用 EventTrigger 播放动画外,还可以将故事板包括在 Resources(资源元素)内,在后台代码中引用故事板的 x:Name,然后使用 Begin 方法播放动画。代码如下:

XAML 代码：
```xml
<UserControl

    xmlns="http://schemas.microsoft.com/winfx/2006/xaml/presentation"
        xmlns:x="http://schemas.microsoft.com/winfx/2006/xaml"
        x:Class="Anim.MainPage"
        Width="640" Height="480">

        <!--将 Storyboard 包含在 Resources 内-->
        <UserControl.Resources>
            <Storyboard x:Name="storyboard1">
                <DoubleAnimation
                    Storyboard.TargetName="ellipse1"
                    Storyboard.TargetProperty="Width"
                    From="150" To="300" Duration="0:0:3" />
            </Storyboard>
        </UserControl.Resources>

        <Canvas Background="AliceBlue">
            <Ellipse x:Name="ellipse1" Fill="GreenYellow" Width="150" Height="200" />
        </Canvas>
</UserControl>
```
[csharp] view plain copy print?
C#代码：
```csharp
namespace Anim
{
    public partial class MainPage : UserControl
    {
        public MainPage()
        {
            //为初始化变量所必须
            InitializeComponent();

            //使用 Begin 方法播放动画
            storyboard1.Begin();
        }
    }
}
```

4.2 Silverlight 线性插值动画

　　线性插值动画支持 DoubleAnimation、ColorAnimation 和 PointAnimaiton 类型的动画。不同类型的动画作用属性的类型也不同，例如 DoubleAnimation 动画的作用属性为 DoubleAnimation 类型，动画的作用属性为 Color 类型，以此类推。这 3 种动画作用类型如下：

(1) DoubleAnimation 是可作用于属性为 Double 类型的 Silverlight 对象的线性插值动画类型,它是最常用的线性插值动画,凡是属于 Double 类型的属性都可以使用它产生线性插值动画效果,如作用于 Ellipse 元素的 Width 属性。

(2) ColorAnimation 是可作用于属性为 Color 类型的 Silverlight 对象的线性插值动画类型,用于改变 Silverlight 对象的填充色调,例如作用于支持 Fill 属性的可见 Silverlight 元素。

(3) PointAnimation 是可作用于属性为 Point 类型的 Silverlight 对象的线性插值动画类型,用于改变某些对象的 X、Y 值,如作用于 EllipseGeometry 元素的 Center 属性。

线性插值动画也称 From/To/By 动画,它是 Silverlight 中最基本的动画对象,From/To/By 可以简单地理解为"开始值/结束值/偏移量"的动画。

动画对象属性的作用如表 4-1 所示。

表 4-1 动画对象属性的作用

名称	描述
From	使动画从 From 属性值到指定的目标对象属性的基础值之间产生动画效果
To	使动画从指定的目标对象属性的基础值到 To 属性指定的值之间产生动画效果
By	使动画从指定的目标对象当前的属性值到 By 属性指定的值之间产生动画效果
Duration	动画执行一次持续的时间长度,Duration 属性格式是"时:分:秒"

除了以上属性以外,线性插值动画还支持对动画播放过程进行控制的属性。

(1) BeginTime:动画开始的时间。默认的单位是天,也可以指定为"时:分:秒"。例如,"0:0:5"等于将 BeginTime 设为 5s。

(2) RepeatBehavior:用来声明动画重复的次数,这个属性支持 3 种类型的值。

① 重复次数:格式为"次数+X",如 1X,2X 表示重复次数。

② 一个时间段:格式为"时:分:秒",如一个动画的 Duration 为 2.5s,将动画的 RepeatBehavior 属性设置为"0:0:5",动画将重复两次。

③ Forever:无限循环。

(3) FillBehavior:决定动画什么时候开始,什么时候结束。可以使动画播放结束后保持到当前位置(HoldEnd),或者结束时恢复到起始位置(Stop)。默认为 HoldEnd。

(4) SpeedRatio:用来增加或减少动画的速度,默认为 1,可以增加或减少。

(5) AutoReverse:播放完成后是否继续向后播放,如果设置为 True,将回到动画开始播放的位置;如果设置为 False,停留到动画播放完的位置。默认值为 False。

4.2.1 DoubleAnimation 动画

DoubleAnimation 动画是最常用的一种线性插值动画,可以用于值为 Double 类型的属性,如图形的长宽等。下面演示一个简单的 DoubleAnimation 动画实例,示例代码如下。

```
<Ellipse Width = "150" Height = "50" Fill = "Orange"
x:Name = "DoubleAnimation">
    <Ellipse.Triggers>
```

```xml
            <EventTrigger RoutedEvent = "Ellipse.Loaded">
                <BeginStoryboard x:Name = "DoubleStoryboard">
                    <Storyboard>
                        <DoubleAnimation
                            Storyboard.TargetName = "DoubleAnimation"
                            Storyboard.TargetProperty = "Height"
                            From = "100" To = "150"
                            RepeatBehavior = "0:0:2"
                            Duration = "0:0:1"
                            SpeedRatio = "2"
                            FillBehavior = "Stop" >
                        </DoubleAnimation>
                    </Storyboard>
                </BeginStoryboard>
            </EventTrigger>
        </Ellipse.Triggers>
</Ellipse>
```

效果是椭圆的高度为 100～150，如图 4-1 所示。

图 4-1 DoubleAnimation 动画

4.2.2 ColorAnimation 动画

当需要使用的对象之间产生 Color（颜色类型）变化上的动画时，可以使用 ColorAnimation 动画。下面的实例演示了一个简单的红绿灯动画，示例代码如下。

```xml
<!--声明资源-->
    <UserControl.Resources>
        <!--声明有红灯变成绿灯故事板 1-->
        <Storyboard x:Name = "Storyboard1">
            <ColorAnimation Storyboard.TargetName = "ellipse1" From = "Red"
                To = "Green"
                Storyboard.TargetProperty =
                "(Ellipse.Fill).(SolidColorBrush.Color)"
                Duration = "0:0:3" />
        </Storyboard>
        <!--声明有绿灯变成红灯故事板 2-->
        <Storyboard x:Name = "Storyboard2">
            <ColorAnimation Storyboard.TargetName = "ellipse2"
                From = "Green" To = "Red"
                Storyboard.TargetProperty =
                "(Ellipse.Fill).(SolidColorBrush.Color)"
                Duration = "0:0:3" />
        </Storyboard>
```

```xml
</UserControl.Resources>
<Canvas x:Name="LayoutRoot" Background="White">
    <!--声明圆角矩形充当背景-->
    <Rectangle Width="314" Height="170"
        Canvas.Left="50"
      Canvas.Top="30"
      RadiusX="20" RadiusY="20"
      Stroke="Green" StrokeThickness="5" />
      <!--红色椭圆形-->
    <Ellipse Width="100" Height="100"
     x:Name="ellipse1" Fill="Red"
     Stroke="Blue" StrokeThickness="5"
    Canvas.Left="77" Canvas.Top="50" />
        <!--绿色椭圆形-->
    <Ellipse Width="100" Height="100"
     x:Name="ellipse2" Fill="Green"
     Stroke="Blue" StrokeThickness="5"
    Canvas.Left="210" Canvas.Top="50" />
     <!--触发故事板1动画-->
    <Button Canvas.Left="90" Canvas.Top="156"
      Content="Red" Height="23"
      Name="btnSwapRed" Click="btnSwapRed_Click" Width="75" />
       <!--触发故事板2动画-->
    <Button Canvas.Left="224" Canvas.Top="156"
      Content="Green" Height="23"
      Name="btnSwapGreen" Click="btnSwapGreen_Click" Width="75" />
</Canvas>
```

运行结果如图4-2所示。

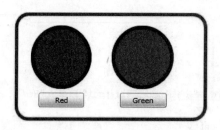

图 4-2　用 ColorAnimation 产生红绿灯动画

上面的示例声明了名为"Ellipse1"和"Ellipse2"的两个椭圆形，并创建了一个 UserControl.Resources 包含两个故事板"Storyboard1""Storyboard2"。两个故事板中都包含了一个 ColorAnimation 元素，其作用分别是将 Fill 属性由红色变为绿色和由绿色变成红色，其中"(Ellipse.Fill).(SolidColorBrush.Color)"属性值表示 Ellipse 对象的 Fill 属性填充的 Brush 集合中的 SolidColorBrush。

4.2.3 PointAnimation 动画

当需要改变的属性类型为 Point 类型时,如 EllipseGeometry 对象的 Center 属性,就需要使用 PointAnimation 对象。下面示例是对 EllipessGeometry 的 Center 属性进行更改,让小球做水平运动,代码如下。

```
<Canvas x:Name="LayoutRoot" Background="White">
    <Path Canvas.Top="200" Canvas.Left="200">
        <Path.Data>
            <!--声明一个几何图形圆 EllipesGeometry1-->
            <EllipseGeometry RadiusX="30" RadiusY="30" Center="0,0"
              x:Name="ellipesGeometry1" />
        </Path.Data>
        <Path.Fill>
            <!--倾斜原点和中心点值都为 0.5-->
            <RadialGradientBrush GradientOrigin="0.5,0.5"
              Center="0.5,0.5">
                <GradientStop Offset="0" Color="#FFECFFDB" />
                <GradientStop Offset="1.0" Color="#FF252825" />
                <GradientStop Offset="0.5" Color="#FF4D6D25" />
            </RadialGradientBrush>
        </Path.Fill>
        <Path.Triggers>
            <!--事件触发器-->
            <EventTrigger RoutedEvent="Path.Loaded">
                <BeginStoryboard>
                    <Storyboard>
                        <PointAnimation From="200,100"
                          AutoReverse="True" To="400,100"
                          Duration="0:0:1"
                          Storyboard.TargetName="ellipesGeometry1"
                          Storyboard.TargetProperty=
                          "Center" RepeatBehavior="Forever" />
                    </Storyboard>
                </BeginStoryboard>
            </EventTrigger>
        </Path.Triggers>
    </Path>
    <TextBlock Text="(From:200,100)" FontSize="15"
Canvas.Left="141" Canvas.Top="210"></TextBlock>
    <TextBlock Text="(To:400,100)"
      FontSize="15" Canvas.Left="400"
      Canvas.Top="210"></TextBlock>
</Canvas>
```

运行结果如图 4-3 所示。

前面已经介绍过如何声明路径和几何椭圆形对象,其中,椭圆形的 Center 是 Point 类型的属性,所以这里使用 PointAnimation 指定 From 和 To 的属性值。

(From:200,100)　　　　　(To:400,100)

图 4-3　通过 PointAnimation 改变图片的位置

这个 PointAnimation 动画中加入了 RepeatBehavior 属性，并声明它的属性值为 Forever，表示动画无限循环，可以为这个动画指定 SpeedRatio 属性值，然后观察动画效果上的变化。

4.3　Silverlight 关键帧动画

关键帧动画与 From/To/By 动画类似，关键帧动画以动画形式显示了目标属性的值。它通过 Duration 创建其目标值之间的过渡。但是，From/To/By 动画创建两个值之间的过渡，而单个关键帧动画可以创建任意数量的目标值之间的过渡。

与 From/To/By 动画不同，关键帧动画没有设置其目标值所需的 From、To 或 By 属性，而是使用关键帧对象描述关键帧动画的目标值。若要指定动画的目标值，需要创建关键帧对象，并将其添加到动画的 KeyFrames 属性。动画运行时，将在指定的帧之间过渡。

若要使用关键帧动画进行动画处理，需要完成下列步骤：

（1）按照对 From/To/By 动画使用的方法声明动画并指定其 Duration。

（2）对于每一个目标值，创建相应类型的关键帧，设置其值和 KeyTime，并将其添加到动画的 KeyFrames 集合内。

（3）按照对 From/To/By 动画使用的方法，将动画与属性相关联。

不同的特性类型有不同的动画类型。若要对采用 Double 的属性（如元素的 Width 属性）进行动画处理，使用生成 Double 值的动画；若要对采用 Point 的属性进行动画处理，使用生成 Point 值的动画，以此类推。

关键帧动画类遵循以下命名约定：

type_ AnimationUsingKeyFrames

其中，type_是要进行动画处理的值的类型。

Silverlight 提供了如表 4-2 所示的关键帧动画类。

表 4-2　关键帧动画类

特 性 类 型	对应的关键帧动画类
Color	ColorAnimationUsingKeyFrames
Double	DoubleAnimationUsingKeyFrames
Point	PointAnimationUsingKeyFrames
Object	ObjectAnimationUsingKeyFrames

每一个关键帧都支持 3 种不同的补间类型,分别为 Linear(线性)、Discrete(离散)、Spline(多键),如表 4-3 所示。

表 4-3 关键帧动画的补间类型

名称	格式	描述
Linear	Linear 类型 KeyFrame	不同类型的 Linear 对象可以使作用目标的属性值在持续时间内产生固定频率渐变,其属性值会随时间的改变而改变,这种变化会实时地反映出来
Discrete	Discrete 类型 KeyFrame	不同类型的 Discrete 对象可以使作用目标的属性值从一个值直接跳到下一个值,在这个过程中不产生任何渐变,所以这种变化只会呈现一个最终结果,不会反映变化的过程
Spline	Spline 类型 KeyFrame	不同类型的 Spline 对象可以使作用目标的属性值产生精确的渐变,通过 KeySpline 属性可以完美模拟更真实的动画

DoubleAnimationUsingKeyFrames、ColorAnimationUsingKeyFrames 和 PointAnimationUsingKeyFrames 类型表示进行动画的"值"的类型,例如 DoubleAnimationUsingKeyFrames 的关键帧对象属性为 Double 类型,动画的补间类型就可以分为 LinearDoubleKeyFrame、DiscreteDoubleKeyFrame 和 SplineDoubleKeyFrame。ColorAnimationUsingKeyFrames 和 PointAnimationUsingKeyFrames 关键帧对象类型可以根据以上命名格式类推。下面分别介绍如何使用 3 种不同类型的关键帧动画。

4.3.1 DoubleAnimationUsingKeyFrames 动画

关键帧动画包含两个重要的属性,即 KeyTime 属性和 Value 属性,它们的作用是在 KeyTime 属性指定的某个时间点对作用目标的 Value 进行控制。简而言之,KeyTime 就是指定在何时到达这个关键帧的 Value。Silverlight 中所有类型的关键帧动画都支持这两个属性。

下面使用 DoubleAnimationUsingKeyFrames 动画实现 3 个椭圆对象的水平运动。

```
XAML 代码:
<UserControl
    xmlns = "http://schemas.microsoft.com/winfx/2006/xaml/presentation"
    xmlns:x = "http://schemas.microsoft.com/winfx/2006/xaml"
    x:Class = "MyDoubleAnimationUsingKeyFrames.MainPage"
    Width = "640" Height = "480">
    <UserControl.Resources>
        <Storyboard x:Name = "LinearStoryboard">
            <!--声明关键帧动画的作用属性为 TranslateTransform 的 x 属性-->
            <DoubleAnimationUsingKeyFrames BeginTime = "00:00:00"
                Storyboard.TargetName = "ellipse1"
                Storyboard.TargetProperty =
                "(UIElement.RenderTransform).(TransformGroup.Children)[0]
                .(TranslateTransform.x)">
                <!--线性补间-->
                <LinearDoubleKeyFrame Value = "500" KeyTime = "00:00:02" />
```

```xml
            </DoubleAnimationUsingKeyFrames>
        </Storyboard>
        <Storyboard x:Name = "DiscreteStoryboard">
            <!-- 声明关键帧动画的作用属性为 TranslateTransform 的 x 属性 -->
            <DoubleAnimationUsingKeyFrames BeginTime = "00:00:00"
                Storyboard.TargetName = "ellipse2"
                Storyboard.TargetProperty =
                "(UIElement.RenderTransform).(TransformGroup.Children)[0]
                .(TranslateTransform.x)">
                <!-- 离散补间 -->
                <DiscreteDoubleKeyFrame Value = "500" KeyTime = "00:00:2.5"/>
            </DoubleAnimationUsingKeyFrames>
        </Storyboard>
        <Storyboard x:Name = "SplineStoryboard">
            <!-- 声明关键帧的作用属性为 TranslateTransform 的 x 属性 -->
            <DoubleAnimationUsingKeyFrames BeginTime = "00:00:00"
                Storyboard.TargetName = "ellipse3"
                Storyboard.TargetProperty =
                "(UIElement.RenderTransform).(TransformGroup.Children)[0]
                .(TranslateTransform.x)">
                <!-- 多键补间 -->
                <SplineDoubleKeyFrame Value = "500" KeyTime = "00:00:4.5"
                    KeySpline = "0.6,0.0,0.9,0.00" />
            </DoubleAnimationUsingKeyFrames>
        </Storyboard>
    </UserControl.Resources>
    <Canvas Background = "White">
        <!-- 声明 ellipse1 -->
        <Ellipse x:Name = "ellipse1" Canvas.Top = "10"
            Width = "100" Height = "100" Fill = "GreenYellow"
            Stroke = "Black" StrokeThickness = "5">
            <Ellipse.RenderTransform>
                <TransformGroup>
                    <TranslateTransform/>
                </TransformGroup>
            </Ellipse.RenderTransform>
        </Ellipse>
        <!-- 声明 ellipse2 -->
        <Ellipse x:Name = "ellipse2" Canvas.Top = "130"
            Width = "100" Height = "100" Fill = "HotPink"
            Stroke = "Black" StrokeThickness = "5">
            <Ellipse.RenderTransform>
                <TransformGroup>
                    <TranslateTransform/>
                </TransformGroup>
            </Ellipse.RenderTransform>
        </Ellipse>
        <!-- 声明 ellipse3 -->
        <Ellipse x:Name = "ellipse3" Canvas.Top = "250"
```

```
                Width = "100" Height = "100" Fill = "Gold"
                Stroke = "Black" StrokeThickness = "5">
                <Ellipse.RenderTransform>
                    <TransformGroup>
                        <TranslateTransform/>
                    </TransformGroup>
                </Ellipse.RenderTransform>
            </Ellipse>
        </Canvas>
    </UserControl>
```

动画效果如图 4-4 所示。

图 4-4　椭圆的水平关键帧动画

在上面的示例中，包含了 3 个椭圆对象，即 ellipse1、ellipse2 和 ellipse3。它们都定义了 TranslateTranform 元素，TranslateTransform 的 X 属性可以用来改变这 3 个椭圆形的水平坐标，使它们从左到右产生一个水平运动动画。它们的 Value 值都是 500，代表它们会从当前的位置向 X 轴运动 500 像素的距离。由于每一组运动使用了不同的补间动画，所以产生了不同的效果。图 4-5 分别是 3 种不同类型补间的运动轨迹。

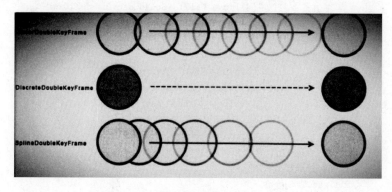

图 4-5　3 种不同类型补间的运动轨迹

示例中分别使用了线性、离散、多键 3 种不同的补间，很显然线性补间产生了一个最基本的水平运动，运动的过程速度是均匀的。离散补间直接从起始点跳到结束点，中间没有任何动画效果。而多键补间是在从起始到结束的过程中产生了一个加速的动画。

4.3.2 ColorAnimationUsingKeyFrames 动画

ColorAnimationUsingKeyFrames 关键帧类型用来操纵 Color 类型属性的变化，通过对 Color 值的改变使作用目标产生色调上的变化。下面使用 ColorAnimationUsingKeyFrames 制作一个可以变色的灯光示例，代码如下：

```xaml
XAML 代码：
<UserControl
    xmlns = "http://schemas.microsoft.com/winfx/2006/xaml/presentation"
    xmlns:x = "http://schemas.microsoft.com/winfx/2006/xaml"
    x:Class = "MyColorAnimationUsingKeyFrames.MainPage"
    Width = "640" Height = "480">
    <Canvas Background = "Gold">

        <!--声明灯光位图-->
        <Image Width = "180" Height = "180" Source = ""
            Canvas.Top = "50" Canvas.Left = "150" />

        <!--声明灯光的变色椭圆形-->
        <Ellipse x:Name = "ellipse1" Width = "100" Height = "100"
            Opacity = "0.5" Fill = "Gray"
            Canvas.Top = "70" Canvas.Left = "200">
            <Ellipse.Triggers>
                <EventTrigger RoutedEvent = "Canvas.Loaded">
                    <BeginStoryboard>
                        <Storyboard x:Name = "ColorStoryboard">
                            <!--声明灯光的变色动画-->
                            <ColorAnimationUsingKeyFrames
                                BeginTime = "00:00:00"
                                Storyboard.TargetName = "ellipse1"
                                Storyboard.TargetProperty =
                                  "(Ellipse.Fill).(SolidColorBrush.Color)">
                                <LinearColorKeyFrame
                                    Value = "Yellow" KeyTime = "00:00:01"/>
                                <LinearColorKeyFrame
                                    Value = "Red" KeyTime = "00:00:02" />
                                <LinearColorKeyFrame
                                    Value = "Green" KeyTime = "00:00:04" />
                            </ColorAnimationUsingKeyFrames>
                        </Storyboard>
                    </BeginStoryboard>
                </EventTrigger>
            </Ellipse.Triggers>
        </Ellipse>
```

```
            <!--声明椭圆对象作为灯光-->
            <Ellipse x:Name="ellipse2" Width="200" Height="200"
                Canvas.Top="20" Canvas.Left="150">
                <Ellipse.Fill>
                    <RadialGradientBrush>
                        <GradientStop Color="#FFFFFFFF" Offset="0"/>
                        <GradientStop Color="#00FFFFFF" Offset="1.0"/>
                    </RadialGradientBrush>
                </Ellipse.Fill>
            </Ellipse>
        </Canvas>
</UserControl>
```

这里使用 ellipse1 作为灯光变色的背景,使用 ellipse2 作为灯泡的灯光效果。ColorAnimationUsingKeyFrames 通过改变 ellipse1 的 Fill 属性产生由黄色到红色再到绿色的渐变效果。KeyTime 属性声明了动画补间产生的时间范围。

4.3.3 PointAnimationUsingKeyFrames 动画

PointAnimationUsingKeyFrames 与 PointAnimation 的作用相似,它们都可以改变 Point 类型的元素属性,不同的是 PointAnimationUsingKeyFrames 可以产生多次位置变化。下面使用 PointAnimationUsingKeyFrames 动画产生一个球体沿着 A、B、C 3 个不同位置产生运动动画,示例代码如下。

```
Html 代码
<UserControl.Resources>
    <Storyboard x:Name="storyboard1" AutoReverse="True" RepeatBehavior="Forever">
        <PointAnimationUsingKeyFrames Storyboard.TargetName="ellipseGeometry1"
Storyboard.TargetProperty="EllipseGeometry.Center" BeginTime="0:0:0">
            <LinearPointKeyFrame Value="75,93" KeyTime="0:0:0"></LinearPointKeyFrame>
            <SplinePointKeyFrame Value="295,260" KeyTime="0:0:2"/>
            <LinearPointKeyFrame Value="525,92" KeyTime="0:0:3"></LinearPointKeyFrame>
        </PointAnimationUsingKeyFrames>
    </Storyboard>
</UserControl.Resources>
<Canvas x:Name="LayoutRoot" Background="White">
    <TextBlock Text="A" FontSize="40" FontFamily="Arial Black" Canvas.Left="58"
Canvas.Top="22" Height="44" Width="32"/>
    <TextBlock Text="B" FontSize="40" FontFamily="Arial Black" Canvas.Left="275"
Canvas.Top="277" Height="44" Width="30"/>
    <TextBlock Text="C" FontSize="40" FontFamily="Arial Black" Canvas.Left="499"
Canvas.Top="22" Height="44" Width="32"/>
    <Rectangle Width="75" Height="10" Fill="Green" Canvas.Left="39" Canvas.Top=
"64"></Rectangle>
    <Rectangle Width="75" Height="10" Fill="Green" Canvas.Left="255" Canvas.Top=
"277"></Rectangle>
```

```
            <Rectangle Width="75" Height="10" Fill="Green" Canvas.Left="482" Canvas.Top=
"64"></Rectangle>
            <!--第一个圆-->
        <Path>
            <Path.Data>
                <EllipseGeometry x:Name="ellipseGeometry1" Center="75,93" RadiusX="20"
RadiusY="20">
                </EllipseGeometry>
            </Path.Data>
            <Path.Fill>
                <RadialGradientBrush Center="0.5,0.5">
                    <GradientStop Offset="0" Color="#FFECFFDB" />
                    <GradientStop Offset="1.0" Color="#FF252825" />
                    <GradientStop Offset="0.5" Color="#FF4D6D25" />
                </RadialGradientBrush>
            </Path.Fill>
        </Path>
            <Button Canvas.Left="255" Canvas.Top="348" Content="运动" Height="23" Name=
"btn_Run" Click="btn_Run_Click" Width="75" />
    </Canvas>
```

上面的示例中使用了一组 Point 关键帧动画并声明其作用属性为 ellipseGeometry 1 元素的 Center 属性,这个关键帧动画中又包含了 3 个 LinegarDoubleKeyFrame 对象,它们声明了 KeyTime 属性 0 秒、2 秒和 4 秒时 Value 属性分别为 3 组坐标(50,50)、(250,230)和 (450,50),这样促使 ellipseGeometry 1 元素的 Center 属性值随着 KeyTime 属性所对应值的改变而改变,从而产生球体 A、B、C 3 点的位置运动,最终完成了从 A 点为起始点,到 B 点后产生反弹,最后到达 C 点的一个球体弹力运动,如图 4-6 所示。

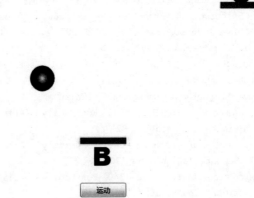

图 4-6　球体的 A、B、C 3 点运动线路

4.4 使用C♯管理动画

前面介绍了如何在 XAML 中使用动画元素和故事板，那么在 C♯ 中是不是也可以使用动画对象和故事板呢？答案是肯定的。下面使用 C♯ 代码创建动画对象和故事板，实现一个椭圆对象的类似放射运动的效果，示例代码如下。

```
XAML:
<UserControl
    xmlns = "http://schemas.microsoft.com/winfx/2006/xaml/presentation"
    xmlns:x = "http://schemas.microsoft.com/winfx/2006/xaml"
    x:Class = "MyStoryBoard.MainPage"
    Width = "640" Height = "480">
    <Canvas x:Name = "parentCanvas" Background = "White"/>
</UserControl>
    C♯:
namespace MyStoryBoard
{
    public partial class MainPage : UserControl
    {
        public MainPage()
        {
            //为初始化变量所必须
            InitializeComponent();

            //创建椭圆对象
            Ellipse ellipse = new Ellipse();
            ellipse.Fill =
                new SolidColorBrush(Color.FromArgb(255,255,0,0));
            ellipse.Width = 150;
            ellipse.Height = 150;

            //添加到 Canvas 中
            parentCanvas.Children.Add(ellipse);
            //创建 Double 动画
            DoubleAnimation anim1 = new DoubleAnimation();
            DoubleAnimation anim2 = new DoubleAnimation();
            DoubleAnimation anim3 = new DoubleAnimation();
            //设置动画的 From 和 To 属性
            anim1.From = 20;
            anim1.To = 400;
            anim2.From = 40;
            anim2.To = 250;
            anim3.From = 30;
            anim3.To = 150;
```

```
            //设置动画时间
            anim1.Duration = new Duration(new TimeSpan(0,0,0,1));
            anim2.Duration = new Duration(new TimeSpan(0,0,0,1));
            anim3.Duration = new Duration(new TimeSpan(0,0,0,1));
            //创建故事板
            Storyboard sb = new Storyboard();
            //设置故事板的时间
            sb.Duration = new Duration(new TimeSpan(0,0,0,1));
            //设置动画是否反复运动,如果设置为 true,椭圆会往复运动
            sb.AutoReverse = false;
            //添加 3 个动画到故事板中
            sb.Children.Add(anim1);
            sb.Children.Add(anim2);
            sb.Children.Add(anim3);
            //设置动画的作用目标
            Storyboard.SetTarget(anim1, ellipse);
            Storyboard.SetTarget(anim2, ellipse);
            Storyboard.SetTarget(anim3, ellipse);
            //设置动画的作用属性
            Storyboard.SetTargetProperty(anim1,
              new PropertyPath("(Canvas.Left)"));
            Storyboard.SetTargetProperty(anim2,
              new PropertyPath("(Canvas.Top)"));
            Storyboard.SetTargetProperty(anim3,
              new PropertyPath("Height"));
            //添加故事板到资源中
            parentCanvas.Resources.Add("storyboard", sb);
            sb.Begin();
        }
    }
}
```

示例的 XAML 代码中只有一个 Canvas 对象,动画内容的椭圆形和动画效果都是在 C#代码中创建的。根据上面的 C#代码,将这个动画的创建过程总结如下。

(1) 创建 DoubleAnimation 对象和设置其属性。

(2) 创建 Storyboard 对象和设置其属性。

(3) 添加 DoubleAnimation 到 Storyboard。

(4) 设置 Storyboard 动画的作用目标和动画的作用属性。

(5) 将 Storyboard 添加到资源字典中。

动画的运行结果如图 4-7 所示。

在 C#中创建 Silverlight 动画的步骤很简单,首先需要创建相应的动画对象和属性,然后将动画和故事板联系起来。一个故事板可以添加多个动画对象。最后与 XAML 一样,需要把故事板加入到资源(Resources)中。

图 4-7 示例动画的运行结果

4.5 Silverlight 多媒体格式与通信协议

Silverlight 提供了强大的影音播放功能。要实现多媒体播放器,开发人员不需要具备很多影音方面的专业知识,也不用考虑客户端的编码解码插件的支持。因为 Silverlight 本身内置了 MediaElement 元素支持影音播放,使用 MediaElement 之前需要了解 MediaElement 支持的格式与通信协议,也便于确定需要播放的视频是否被 Silverlight 支持。

4.5.1 MediaElement 支持与不支持的视频和音频格式

1. 支持的视频格式

1) 原始视频
2) RGBA 格式
(1) 未压缩的 32 位 Alpha、红色、绿色、蓝色。
(2) 在 Windows Phone 7 中,Alpha 通道被忽略了。
3) YV12 格式:YCrCb(4:2:0)
(1) 未压缩的 YCrCb(4:2:0)。
(2) 在 Windows Phone 7 中不受支持。
4) RGBA:32 位 Alpha、红色、绿色、蓝色
5) Windows Media Video 和 VC-1 格式 WMV1:Windows Media Video 7
(1) 支持简单配置文件、主配置文件和高级配置文件。
(2) 仅支持渐进式(逐行扫描)内容。
6) WMV2:Windows Media Video 8
7) WMV3:Windows Media Video 9
(1) 支持简单配置文件和主配置文件。
(2) 仅支持渐进式(逐行扫描)内容。
8) WMVA:Windows Media 视频高级配置文件,非 VC-1
9) WVC1:Windows Media 视频高级配置文件,VC-1
(1) 支持高级配置文件。
(2) 仅支持渐进式(逐行扫描)内容。
10) H264 (ITU-T H.264 / ISO MPEG-4 AVC)格式
(1) 支持 H.264 和 MP43 编解码器。
(2) 支持基本配置文件、主配置文件和高配置文件。
(3) Windows Phone 7 支持高达 3.0 级。
请注意 3.0 级定义 30 帧/秒的最大支持的分辨率为 720×480。
其他帧速率有不同最大支持的分辨率。
(4) 仅支持渐进式(逐行扫描)内容。
(5) 仅支持 4:2:0 色度二次采样配置文件。
(6) 桌面 Silverlight 仅支持附件 B 宇空实验室格式(即启动代码)。

注意：媒体库，如 SmoothStreamingMediaElement，通过编程方式将 AVC NALs 转换为附件 B NALs 支持 AVC 宇空实验室格式。因为 Silverlight 支持 H.264 示例的部分加密，如果未加密 NAL 标头，则此转换也可能出现在加密内容上。MPEG 4 第十部分描述了附件 B 宇空实验室格式。

（7）支持具有 Mp4 的 PlayReady DRM（H264 和 AAC-LC）。

11）H.263 格式

（1）Silverlight 的桌面版本不支持。

（2）在 Windows Phone 7 中不受 MediaStreamSource 类支持。

12）MPEG-4 第二部分格式

支持简单配置文件和高级配置文件。

2．支持的音频格式

1）PCM / WAV 格式"1"。这是线性 8 位或 16 位脉冲编码调制。大致上说，这是 WAV 格式。

这是线性 8 位或 16 位脉冲编码调制。大致来说，这是 WAV 格式。

2）Microsoft Windows Media Audio Standard formats。"353"为 Microsoft Windows Media Audio v7、v8 和 v9.x Standard（WMA Standard）

3）Microsoft Windows 媒介音频专业模式。"354"为 Media Audio v9.x and v10 专业版（WMA 专业版）

（1）支持 32～96Kb/s 范围内的 WMA 10 Professional 低比特率（LBR）模式的全保真解码。

（2）多声道（5.1 和 7.1 环绕）音频内容自动混缩为立体声。

（3）24 位音频将返回静音。

（4）采样率超过 48000 将在同域时返回无效格式错误代码，在跨域时返回 4001。

4）MP3 "85"：（ISO MPEG 1 层Ⅲ）格式（MP3）。

5）AAC "255"：（ISO 高级音频编码）（AAC）格式。

（1）支持达到全保真的低复杂度（AAC-LC）解码（最高 48kHz）。

（2）高效（HE-AAC）编码内容。将仅解码为半保真（最高 24kHz）。

① HE-AAC v1（AAC+）和支持的 HE-AAC v2（eAAC+）。

② 桌面 Silverlight 将只在半保真解码（最多 24kHz）。

（3）不支持多声道（5.1 环绕）音频内容。

6）AMR-NB（自适应多速率窄带）格式

（1）Silverlight 的桌面版本不支持。

（2）不受 MediaStreamSource 类支持。

3．不支持的格式

MediaElement 虽然支持众多的多媒体格式，但是它不是万能的，也有不支持的编解码器格式，如下：

➢ Windows Media 屏幕

➢ Windows Media 专业无损音频

➢ Windows Media 语音

➢ Windows Phone 7 不支持动态变化的视频帧

➢ 隔行扫描的视频内容
➢ Windows Media 视频和 MP3 的组合（WMV 视频＋MP3 音频）

使用奇数（不能被 2 整除）维度的帧的 Windows Media 视频，例如 127×135。

4.5.2　MediaElement 的媒体播放机制

MediaElement 支持在 Windows Media Server 上及时播放，并且可以适应选择的播放。如果对 MediaElement 的播放地址指定一个 MMS 地址，则会优先使用流媒体方式播放。

如果不能够获取流数据，MediaElement 会使用渐进下载播放，即下载一部分播放一部分的方式。下载的内容会保存在浏览器的缓存区。当再次播放相同视频文件时不会再重新下载视频文件，MediaElement 直接播放缓存中已下载的多媒体文件；反之，如果指定 URI 为 HTTP 或 HTTPS，在不成功的情况下 MediaElement 就会尝试使用 MMS。此外，MediaElement 还支持 ASX 的播放列表。

在 Silverlight 中的流播放和渐进下载播放最终都是使用 HTTP 协议来播放，如果使用 MediaElement 播放 Windows Media Serve 的 MMS 视频，那么一定要打开 MMS 的 HTTP 流播放协议支持，否则 MediaElement 将无法正常播放 MMS 流，RTSP 也是如此。

H.264 视频格式是 MediaElement 一开始支持的视频格式，H.264 是一种高清视频格式，但是它不具有明显的视频流特征，所以 H.264 视频和 AAC 音频格式只支持 Progressive，下载 Smooth 流和 Adaptive 流（需要使用 Expression Encoder 2 发布），AAC 格式的音频最高支持 48kHz 的采样。

使用 Live Smooth Streaming 需要先安装 IIS Live Streaming server，添加发布点等工作。详细的操作方法这里不做介绍，有兴趣的读者可以参考 http:/Imsdn.microsoft.com/zh-cn/dd767539.upx 的介绍。

4.6　MediaElement 对象

MediaElement 是一个可以在其控件面板上播放视频或音频的控件，它属于 UI 元素，是 UIElement 对象的子类。MediaElement 具有 UIElement 具有的属性，例如可以设置透明度，或者设置图像裁切效果。而且还支持输入操作，可以获得输入焦点，通过 Height 和 Width 属性可以设置视频显示的高度和宽度，以及为 MediaElement 对象设置变换效果等。

MediaElement 控件的基本用法非常容易上手，当学过如何使用该控件以后，会发现它还提供许多高级功能。现在，循序渐进地学习该控件的用法，第一步是看看如何使用 MediaElement 控件完成一些最常见的任务。

MediaElement 控件加入页面中，并设置其 Source 属性为想播放的视频文件所在的

URL 地址。如下所示：

```
<Canvas xmlns = "http://schemas.microsoft.com/client/2007"
xmlns:x = "http://schemas.microsoft.com/winfx/2006/xaml"
"Background = "White"
>
    <MediaElement Source = "balls.wmv"/>
</Canvas>
```

这样将会自动地载入并播放视频文件。该视频的尺寸由以下规则决定：

（1）通过设置 Width 属性和 Height 属性，可以指定 MediaElement 控件的宽和高。

（2）只设置 Width 属性或 Height 属性，MediaElement 控件将会按比例拉伸视频尺寸。

（3）Width 属性和 Height 属性都没有设置，MediaElement 控件就会按照视频原来的尺寸显示视频。如果视频尺寸超出了 Silverlight 控件的可见区域，MediaElement 会裁剪掉超出的部分。

下面看一个示例。本章中用的 balls.wmv 视频大小为 480×360。如果使用 MediaElement 控件显示该视频，并且不指定其宽和高，那么视频将会以 480×360 的大小播放。如果 Silverlight 控件大小为 200×200，那么只能看到该视频左上角边长为 200 像素的正方形区域。图 4-8 显示了视频被切割的情况。

图 4-8 视频被切割效果

正如前面讲到的一样，在视频播放时，MediaElement 控件的大小是很重要的。如果没有指定 MediaElement 控件的大小，而视频尺寸又超过了 Silverlight 控件的大小，那么视频将会被剪切。

要控制 MediaElement 控件的高和宽，需要使用它的 Height 和 Width 属性。当控件显示时，视频文件将会被拉伸（或收缩）到媒体控件的大小。如果媒体控件的大小超过了 Silverlight 控件的大小，那么媒体控件将会被裁减到 Silverlight 控件的大小。

4.7 视频拉伸模式

除了通过使用 Height 和 Width 硬性改变视频的大小之外，MediaElement 还支持拉伸模式，它是通过使用 MediaElement 控件的 Stretch 属性改变这些显示效果。该属性可以有以下 4 个不同的值。

(1) None：不会拉伸。如果 MediaElement 控件大于视频实际尺寸，视频将位于控件的中间。如果 MediaElement 控件小于视频尺寸，将只显示视频的中间部分。例如，视频的尺寸为 480×360，如果 MediaElement 控件的尺寸设为 200×200，并且 Stretch 设置为 None，那么视频中间 200×200 的区域将会被显示。

(2) Uniform：默认拉伸模式，该模式下将维持视频的长宽比例并给视频添加边框。

(3) UniformToFill：该模式下会按比例拉伸视频并剪切视频适应窗口的大小。例如，如果视频的宽度大于高度（如，480×360）并且拉伸适应 200×200 大小的窗口，视频的边将会被切掉以适应可视窗口的大小。

(4) Fill：该模式下将使用视频填充 MediaElement 控件。

4.8 MediaElement 状态管理

MediaElement 在播放多媒体文件的过程中会返回播放的状态 CurrentState 和一个状态时间 CurrentStateChanged 事件，通过它们可以监听 MediaElement 对象的实时状态。

CurrentState 状态如下。

(1) Buffering：缓冲区未满，此时正在加载要播放的媒体。在此状态中，Position（播放的位置）不能够向前。

(2) Closed：不包含媒体，媒体已关闭。

(3) Opening：正在尝试打开 Source 属性指定的 URI，将开始进行缓冲或下载。

(4) Paused：暂停播放。

(5) Playing：正在播放。

(6) Stopped：包含媒体地址，可能是未开始播放或已经停止播放。

4.9 缓冲进度和下载进度

MediaElement 对象在实际应用中需要考虑网络环境是否通畅等因素，为了提高用户体验，可考虑在适当的位置使用 BufferingProgress（缓冲进度）属性/ BufferingProgressChanged 事件和 DownloadProgress（下载进度）属性/DownloadProgressChanged 事件。

DownloadProgress 代表多媒体内容下载的进度，通常是在使用 HTTP 或 HTTPS 的 URI 显示内容的，一般当渐进式的下载不可用或不支持时会要求下载完整个多媒体资源后再能播放。代码见下面的例子。

```
XAML:
<Grid x:Name = "LayoutRoot" Background = "White">
    <!-- 多媒体对象 -->
    <MediaElement x:Name = "media"
        Source = "/Video/SilverlightIntro.wmv"
```

```
                DownloadProgressChanged = "media_DownloadProgressChanged"/>
        <!-- 显示下载进度 -->
        <TextBlock x:Name = "tbk"
            FontSize = "15"
            Width = "120" Height = "50"
            Text = "Download:0%">
</Grid>
C#
private void media_DownloadProgressChanged(object sender,
 RoutedEventArgs e)
{
    //计算百分比
    double downVal = media.DownloadProgress * 100;
    tbk.Text = "Download:" + downVal.ToString();
}
```

BufferingProgress 代表多媒体缓冲值,介于 0~1 之间,显然可以将它的值乘以 100 表示缓冲的百分比值。对于一般的多媒体资源,当缓冲到 5% 或完成缓冲后就会引发 BufferingProgressChanged 事件,在事件中,就可以获取当前的缓冲值(因为 MediaElement 只有当缓冲满时才会播放多媒体文件)。另外还可以手动设置缓冲的时间,即设置 BufferingTime 属性(标准时间格式),当网络条件不好时,适当调整缓冲时间可以减少视频音频的暂停时间。见下面的例子。

```
XAML:
<Grid x:Name = "LayoutRoot" Background = "White">
    <!-- 多媒体对象 -->
    <MediaElement x:Name = "media"
        Source = "/Video/SilverlightIntro.wmv"
        BufferingProgressChanged = "media_BufferingProgressChanged" />
    <!-- 显示下载进度 -->
    <TextBlock x:Name = "tbk"
        FontSize = "15"
        Width = "120" Height = "50"
        Text = "Buffering:0%">
</Grid>
C#
private void media_BufferingProgressChanged(object sender,
 RoutedEventArgs e)
{
    //计算百分比
    double bufferVal = media.BufferingProgress * 100;
    tbk.Text = "Buffering: " + bufferVal.ToString();
}
```

另外还可以手动设置缓冲的时间 BufferingTime 属性,当网络条件不佳时,适当调整缓冲时间可以减少视频音频的暂停时间,BufferingTime 是一个标准的时间格式,例如 00:00:30 是设置一个 30s 的缓冲时间。

4.10 获取和控制播放位置

在播放多媒体资源时，MediaElement 除了可以返回播放状态、下载和缓冲外，还可以通过 NaturalDuration 和 Position 获取播放资源的位置，Position 支持 set 访问，所以可以用其控制视频和音频的播放位置。

```csharp
C#
public VideoElementF()
{
    InitializeComponent();
    //显示播放的时分秒格式
    tbk.Text = string.Format("{0}:{1}:{2}",
        media.Position.Hours < 10 ? "0"
        + media.Position.Hours.ToString():
        media.Position.Hours.ToString(),
        media.Position.Minutes < 10 ? "0"
        + media.Position.Minutes.ToString():
        media.Position.Minutes.ToString(),
        media.Position.Seconds < 10 ? "0"
        + media.Position.Seconds.ToString():
        media.Position.Seconds.ToString());
    //多媒体资源的总播放长度
    double totalTime = media.Position.TotalSeconds;
    //多媒体资源的当前播放长度
    double currentTime = media.NaturalDuration.TimeSpan.TotalSeconds;
    //重新设置 MediaElement 的播放位置
    media.Position = TimeSpan.FromSeconds(30);
}
```

4.11 视频画刷的应用

VideoBrush 是一种类似于 LinearGradientBrush 或 ImageBrush 的 Brush 对象。但是，VideoBrush 使用一个视频文件绘制一个图形区域而不是渐变或图像绘制区域。所以需要事先创建一个 MediaElement 对象，然后使用 VideoBrush 的 SourceName 属性指定视频的提供者，就是 MediaElement 的名字。下面使用 VideoBrush 分别填充一个矩形和一个椭圆形，代码如下：

```xml
<!-- 完全透明的 MediaElement 对象 -->
<MediaElement x:Name = "media"
              Source = "/Video/SilverlightIntro.wmv"
```

```
                    Opacity = "0">
<!-- 视频画刷的填充对象 -->
<Ellipse Width = "500" Height = "400"
         Stroke = "Red"
         StrokeThickness = "5">
<Ellipse.Fill>
<!-- 视频画刷 -->
<VideoBrush SourceName = Mmedia" Stretch = "UniformToFill"/>
</Ellipse.Fill>
</Ellipse>
```

总 结

本章主要介绍如何使用 Silverlight 中的多媒体对象 MediaElement。通过制作全功能的视频播放器介绍如何实现对 MediaElement 各项目属性与事件的应用。Silverlight 的多媒体播放能力是自 Silverlight 1.0 推出以来的重要特性,不但可以把它作为一个单纯的视频播放器,还可以使用 VideoBrush 把一个视频作为填充元素应用在任何支持它的对象上面。

动画是 RIA 技术的关键能力,动画也是在用户交互过程中 Silverlight 与 Ajax 技术有本质区别的明显标志,在应用程序的交互功能中适当地使用动画效果可以提高用户的使用兴趣。

通过本章的学习,可以发现 Silverlight 给开发者提供了更简洁的动画设计途径,使普通的程序员在不依赖设计师的情况下就可以快速地创建简单、逼真的 Silverlight 动画效果,为用户带来上乘的 UI 体验。

作 业

1. 简述触发器的几个重要属性。各自作用是什么?
2. 结合前面学习的知识,制作一个图片模糊的效果动画。通过 C# 代码实现控制动画,通过触发器控制动画,效果如图 4-9 所示。

图 4-9 图片模糊效果

第5章

Silverlight 与 HTML、JavaScript 三者交互

学习目标

- ➢ 了解 Object 标记的参数
- ➢ 掌握 Silverlight 中获取初始化/网页参数
- ➢ 掌握 Silverlight 与 HTML 交互
- ➢ 掌握 Silverlight 与 JavaScript 交互
- ➢ 掌握 Silverlight 与 ASP.NET 集成

5.1 Silverlight 对象模型与 DOM

HTML DOM 意为 HTML 文档对象模型，Silverlight 的对象模型是以 HTML 页面或 ASPX、PHP 等网页为宿主而运行的。

HTML 可以由浏览器直接解析并将内容呈现出来，但是 Silverlight 的 XAML 内容（即下载的 XAP 文件）不能被浏览器直接解释，它需要通过 Silverlight 插件的 XAML 分析器分析并运行，最后与 HTML 页面一样呈现在客户端浏览器中。

Silverlight 的跨浏览器能力是有针对性的，对于 Internet Explorer 浏览器，使用的是 ActiveX 插件模式，而对其他浏览器，则使用 Netscape API 插件模式。

在一个网页内，Silverlight 的内容插件被包含基于一个 DIV 的 Object 元素中，这个 DIV 元素作为 Silverlight 应用程序插件容器，这样使 Silverlight 应用程序可以被部署并运行于网页中的任何一个位置，在 HTML 页面中承载 Silverlight 应用程序的基本代码如下。

```
XAML:
< object data = "data:application/x - silverlight - 2,"
    type = "application/x - silverlight - 2" width = "300" height = "300">
```

```
            < param name = "source" value = "ClientBin/Sample.xap"/>
</object >
```

其中,width 和 height 用来声明 Silverlight 应用程序的显示的尺寸,如果指定的尺寸小于 Silverlight 应用程序中的 UserControl 尺寸,那么超出部分的内容会被裁剪掉。默认情况下,一个标准的 Silverlight Object 将包含以下代码。

```
XAML:
< object data = "data:application/x-silverlight-2,
    " type = "application/x-silverlight-2"
        width = "100%" height = "100%">
        < param name = "source" value = "ClientBin/Sample.xap"/>
        < param name = "onError" value = "onSilverlightError"/>
        < param name = "background" value = "white" />
        < param name = "minRuntimeVersion" value = "3.0.40624.0"/>
        < param name = "autoUpgrade" value = "true"/>
            < a href = "http://go.microsoft.com/fwlink/? LinkID = 149156&v
                = 3.0.40624.0" style = "text-decoration:none">
                < img src = "http://go.microsoft.com/fwlink/? LinkId
                    = 108181" alt = "Get Microsoft Silverlight"
                    style = "border-style:none"/>
            </a>
</object >
```

在这个 Object 标记中包括很多属性,这些属性声明了 Silverlight 插件运行加载时的相关资料,例如尺寸、应用程序文件、版本等。表 5-1 中列出了以上代码包含的属性的具体作用。

表 5-1 Silverlight 插件的基本属性

属 性 名 称	描 述
Source	指定 Silverlight 应用程序包(XAP)文件,它可以是一个相对路径,也可以是一个 URI
type	Silverlight 插件基本标识。默认值:application/x-silverlight-2,如果使用 Silverlight 3.0 SDK 创建的 Silverlight 应用程序,那么它的值就是相应的 application/x-silverlight-3
data	设置该属性的值有助于防止某些浏览器性能下降。 默认值为 data:application/x-silverlight
width	指定页面中 Silverlight 插件区域的初始宽度,可以设置为像素值或百分比
height	指定页面中 Silverlight 插件区域的初始高度,可以设置为像素值或百分比
background	声明 Silverlight 插件的背景色
minRuntimeVersion	设置最低运行的 Silverlight 运行时版本号。autoUpgrade 设置是否允许 Silverlight 进行自动更新
initParams	允许自定义 Silverlight 插件初始化参数。这些参数在 Silverlight 应用程序初始化时将被传递给 Silverlight 启动事件
onerror	通过 JavaScript 显示 Silverlight 的错误信息
enableHtmlAccess	允许 Silverlight 访问网页中的 JavaScript 和 DOM,默认值为 true

续表

属 性 名 称	描 述
airHtml	Silverlight 插件安装提示信息,格式为 HTML
EnableGPUAcceleration	设置 Silverlight 3 是否启用 GPU 加速,GPU 加速在 Silverlight 3 中是默认关闭的

除了上面描述的属性之外,Silverlight 还支持自定义安装信息和 JavaScript 错误信息提示。

5.2 获取 Silverlight 插件的错误信息

onerror 属性可以指定一个 JavaScript 函数返回 Silverlight 的错误信息,默认函数名为 onSilverlightError。可以通过 onSilverlightError 函数获取错误信息,以调试 Silverlight 应用程序。onSilverlightError 函数的内容如下:

```
HTML:
< script type = "text/javascript">
        function onSilverlightError(sender, args) {
            var appSource = "";
            if (sender != null && sender != 0) {
                appSource = sender.getHost().Source;
            }
            var errorType = args.ErrorType;
            var iErrorCode = args.ErrorCode;
            if (errorType == "ImageError" || errorType == "MediaError")
            {
                return;
            }
            var errMsg = "Unhandled Error in Silverlight Application"
               + appSource + "\n";
            errMsg += "Code: " + iErrorCode + " \n";
            errMsg += "Category: " + errorType + " \n";
            errMsg += "Message: " + args.ErrorMessage + " \n";
          if (errorType == "ParserError")
          {
            errMsg += "File: " + args.xamlFile + " \n";
            errMsg += "Line: " + args.lineNumber + " \n";
            errMsg += "Position: " + args.charPosition + " \n";
        } else if (errorType == "RuntimeError")
        {
            if (args.lineNumber != 0) {
                errMsg += "Line: " + args.lineNumber
                   + " \n";
                errMsg += "Position: " + args.charPosition
                   + " \n";
```

```
                    }
                    errMsg += "MethodName: " + args.methodName
                    + "\n";
                }
            throw new Error(errMsg);
        }
</script>
```

可以修改上面的函数内容,使错误提示更为友好,其中 errorType(错误类型)分为 ParserError(转换错误)和 RuntimeError(运行时错误),一般在部署环境中不包含它。

5.3 在 Silverlight 中获取初始化参数和网页参数

一个可复用的 Silverlight 应用程序需要根据 HTML 页面传递的不同参数显示不同的内容,Silverlight 提供了两种获取 HTML 页面传递过来参数的方式,它们分别是 initParams 和 HtmlDocumentQueryString。

(1) initParams:设置 Silverlight 的初始化参数值。可以在 Silverlight 应用程序运行后在 ApplicationStartup 事件中通过 StartupEventArgs 或 App.Current.Host.InitParams 获取参数值。

(2) HtmlDocument.QueryString:直接获取 HTML 页面上的参数值。可以在 Silverlight 程序的任意位置调用。

下面通过示例介绍通过设置参数并分别使用 APP 和 HtmlPage 获取参数,代码如下。

```
HTML:
<object data="data:application/x-silverlight-2,"
type="application/x-silverlight-2" width="100%" height="100%">
    <param name="source" value="ClientBin/Lession5Demo.xap"/>
    <param name="onError" value="onSilverlightError" />
    <param name="background" value="white" />
    <param name="minRuntimeVersion" value="4.0.50826.0" />
    <param name="autoUpgrade" value="true" />
    <param name="initParams" value="id=a,name=a"/>
    <a href="http://go.microsoft.com/fwlink/?LinkID=149156&v
      =4.0.50826.0" style="text-decoration:none">
        <img src="http://go.microsoft.com/fwlink/?LinkId=161376"
          alt="获取 Microsoft Silverlight" style="border-style:none"/>
    </a>
</object>
C#:
private void Application_Startup(object sender, StartupEventArgs e)
    {
        string id = e.InitParams["id"].ToString();
        string name = e.InitParams["name"].ToString();
        this.RootVisual = new MainPage(id,name);
```

```
                }
//在 MainPage 页的构造函数中获取参数
    public MainPage(string id, string name)
        {
            InitializeComponent();                //通过 App 启动项获取
            //通过当前主程序
            this.tbApp.Text = "id:" + id + " name:" + name;
            //通过页面参数
            this.tbHost.Text =
                    "id:" + App.Current.Host.InitParams["id"]
                    + " name:"
                    + App.Current.Host.InitParams["name"];
            this.tbURL.Text =
                    "id:" + HtmlPage.Document.QueryString["id1"]
                    + " name:"
                    + HtmlPage.Document.QueryString["name1"];
        }
```

运行结果如图 5-1 所示。

从App.xaml.cs文件传递的参数：	id:a	name:a
从Host中获取：	id:a	name:a
从URL中获取：	id:a	name:a

图 5-1　运行结果

5.4　Silverlight 捕获浏览器信息

Silverlight 与浏览器相关功能都被封装在 System.Windows.Browser 命名空间中，在 C#代码中引用它可以在 Silverlight 中获取浏览器的相关信息。

BrowserInformation 类中包括了大部分浏览器相关信息，这里设计一个简单获取浏览器信息的示例，代码如下。

```
XAML:
<UserControl.Resources>
        <Style x:Key = "txtStyle" TargetType = "TextBlock">
            <Setter Property = "FontSize" Value = "20"/>
            <Setter Property = "Margin" Value = "10"/>
            <Setter Property = "TextWrapping" Value = "Wrap"/>
        </Style>
</UserControl.Resources>
<StackPanel Background = "White">
        <TextBlock x:Name = "tbkName"
              Style = "{StaticResource txtStyle}"/>
        <TextBlock x:Name = "tbkVer"
```

```
                Style = "{StaticResource txtStyle}"/>
            <TextBlock x:Name = "tbkPlatform"
                Style = "{StaticResource txtStyle}"/>
            <TextBlock x:Name = "tbkUserAgent"
                Style = "{StaticResource txtStyle}"/>
</StackPanel>
    C#:
public htmldom()
        {
            InitializeComponent();            //取得浏览器信息对象
            //请先引用 System.Windows.Browser;
            BrowserInformation browserInfo
                = HtmlPage.BrowserInformation;
            tbkName.Text = string.Format("Name: {0}"
                , browserInfo.Name);
            tbkVer.Text = string.Format("BrowserVersion: {0}"
                , browserInfo.BrowserVersion);
            tbkPlatform.Text = string.Format("Platform: {0}"
                , browserInfo.Platform);
            tbkUserAgent.Text = string.Format("UserAgent: {0}"
                , browserInfo.UserAgent);
        }
```

运行结果如图 5-2 所示。

Name: Microsoft Internet Explorer
BrowserVersion: 5.0
Platform: Win32
UserAgent: Mozilla/5.0 (compatible; MSIE 9.0; Windows NT 6.1; WOW64; Trident/5.0; SLCC2; .NET CLR 2.0.50727; .NET CLR 3.5.30729; .NET CLR 3.0.30729; Media Center PC 6.0; InfoPath.3; .NET4.0C; .NET4.0E)

图 5-2 Silverlight 捕获浏览器信息运行结果

5.5 Silverlight 操作 HTML 元素

Silverlight 提供了 HtmlPage 和 HtmlElement，分别代表 HTML 页面和 HTML 元素，例如<img/>、<div/>、<table/>，这些都是基本的 HTML 元素，在 Silverlight 中，可以轻松地与它们打交道，这里实现通过 Silverlight 的文件框动态获取一个 HTML 的 img 元素并改变它的尺寸，示例如下。

```
XAML:
<UserControl.Resources>
        <Style x:Key = "textBox" TargetType = "TextBox">
```

```xml
            <Setter Property="Width" Value="200"/>
<Setter Property="Height" Value="35"/>
            <Setter Property="FontSize" Value="20"/>
            <Setter Property="HorizontalAlignment" Value="Left"/>
          <Setter Property="VerticalAlignment" Value="Center"/>
  </Style>
    </UserControl.Resources>
    <Grid Background="GreenYellow" ShowGridLines="True">
        <Grid.RowDefinitions>
            <RowDefinition/>
            <RowDefinition/>
            <RowDefinition/>
        </Grid.RowDefinitions>
        <Grid.ColumnDefinitions>
         <ColumnDefinition Width="150"/>
         <ColumnDefinition Width="250"/>
        </Grid.ColumnDefinitions>
        <TextBlock Text="图片宽度:" FontSize="25"
                HorizontalAlignment="Right"
                VerticalAlignment="Center"
              Grid.Row="0" Grid.Column="0"/>
        <TextBox x:Name="textBox1"
            Grid.Row="0" Grid.Column="1"
              Style="{StaticResource textBox}"/>
        <TextBlock Text="图片高度:" FontSize="25"
                HorizontalAlignment="Right"
                VerticalAlignment="Center"
              Grid.Row="1" Grid.Column="0"/>
        <TextBox x:Name="textBox2"
            Grid.Row="1" Grid.Column="1"
              Style="{StaticResource textBox}"/>
        <Button x:Name="Button1" Width="150" Height="50"
            Grid.Row="2" Grid.Column="1"
             FontSize="30" Content="设 置"
            Click="Button1_Click"/>
</Grid>
```

C#：

```csharp
private void Button1_Click(object sender, RoutedEventArgs e)
        {
            //获取 HTML 中的 img1 对象
            HtmlElement img =
                HtmlPage.Document.GetElementById("img1");
//根据输入值设置 img1 的 HTML 属性 Width 和 Height
img.SetAttribute("Width", textBox1.Text);
            img.SetAttribute("Height", textBox2.Text);
        }
```

运行结果如图 5-3 所示。

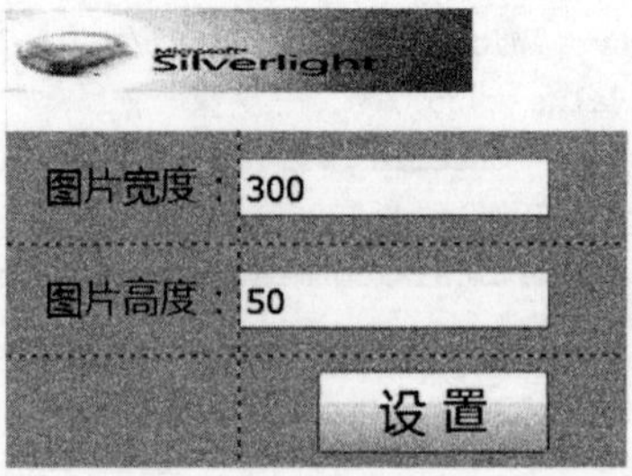

图 5-3 动态获取 HTML 的 img 元素并改变其尺寸

5.6　HTML 元素操作 Silverlight 对象

不仅可以在 Silverlight 中获取和操作 HTML 对象，在 HTML 中同样可以将事件通知给 Silverlight 应用程序，并反映事件的变化。下面通过 HTML 的文本框元素动态地改变 Silverlight 中椭圆形的背景色，示例如下。

```
XAML:
<Grid x:Name = "LayoutRoot" Background = "GreenYellow">
        <!-- 椭圆形对象,用户显示 HTML 设置的颜色 -->
        <Ellipse x:Name = "ellipse1"
                Stroke = "Black" StrokeThickness = "3"
                Width = "150" Height = "150" Fill = "Orange"/>
        <!-- 用来显示颜色值 -->
        <TextBlock x:Name = "textBlock1" Text = "Ellipse 默认值"
                Foreground = "White" FontSize = "26"/>
</Grid>
    C#:
public htmldom2()
{
        InitializeComponent();              //获取 select 对象
       //添加 select 的 onchange 事件
       HtmlElement select = HtmlPage.Document.GetElementById("sel");
       select.AttachEvent("onchange",
             new EventHandler<HtmlEventArgs>(select_onChange));
}
public void select_onChange(object sender, HtmlEventArgs e)
{
         //获取 select 的选择值
         HtmlElement select = sender as HtmlElement;
         string value = select.GetAttribute("value");
         textBlock1.Text = value;
         //根据用户选择值改变椭圆形的 Fill 值
         switch (value){
            case "Red":
```

```
                ellipse1.Fill = new SolidColorBrush(Colors.Red);
                break;
            case "Green":
                ellipse1.Fill = new SolidColorBrush(Colors.Green);
                break;
            case "Blue":
                ellipse1.Fill = new SolidColorBrush(Colors.Blue);
                break;
        }
    }
```

运行结果如图 5-4 所示。

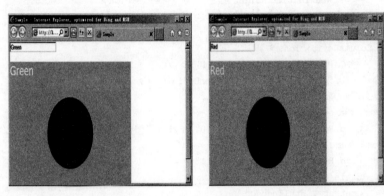

图 5-4　通过 HTML 的文本框元素动态地改变 Silverlight 中椭圆形的背景色结果

5.6.1　使用 HttpUtility 类

Silverlight 提供了 HttpUtility 类,可以实现对 URL 的编码和解码(UrlEncode/UrlDecode 及 HtmlEncode/HtmlDecode)。这里×××-Encode 和×××-Decode 是成对使用的。

1. UrlEncode 和 UrlDecode

在网页程序开发过程中,页面之间常使用 Get 和 Post 方式传递参数数据,当使用中文作为网页参数传递时,常存在一些问题,通常的解决方法是使用 UrlEncode 和 UrlDecode,把中文字符编码后传递,然后再解码。

2. HtmlEncode 和 HtmlDecode

当网页中需要保存用户输入的含有 HTML 标记的内容时,基于安全性考虑,常需要将其进行 HtmlEncode 编码,在显示这些 HTML 时,通过 HtmlDecode　解码后显示出来,代码如下。

```
XAML:
<UserControl.Resources>
    <Style x:Key="textBox" TargetType="TextBox">
        <Setter Property="Width" Value="350"/>
        <Setter Property="Height" Value="40"/>
```

```xml
            <Setter Property="FontSize" Value="20"/>
        </Style>
        <Style x:Key="button" TargetType="Button">
            <Setter Property="Width" Value="150"/>
            <Setter Property="Height" Value="40"/>
            <Setter Property="FontSize" Value="15"/>
        </Style>
    </UserControl.Resources>
    <Grid x:Name="LayoutRoot" Background="GreenYellow">
        <Grid.RowDefinitions>
            <RowDefinition Height="80"></RowDefinition>
            <RowDefinition Height="80"></RowDefinition>
            <RowDefinition Height="80"></RowDefinition>
            <RowDefinition Height="80"></RowDefinition>
        </Grid.RowDefinitions>
        <Grid.ColumnDefinitions>
            <ColumnDefinition Width="400"></ColumnDefinition>
            <ColumnDefinition Width="200"></ColumnDefinition>
        </Grid.ColumnDefinitions>
        <TextBox x:Name="textBlock1"
            Grid.Row="0" Grid.Column="0" Style="{StaticResource textBox}"/>
        <Button x:Name="btn_HtmlEncode"
            Grid.Row="0" Grid.Column="1"
            Style="{StaticResource button}"
            Content="HtmlEncode"
            Click="btn_HtmlEncode_Click"/>
        <TextBox x:Name="textBlock2"
            Grid.Row="1" Grid.Column="0"
            Style="{StaticResource textBox}"/>
        <Button x:Name="btn_HtmlDecode"
            Grid.Row="1" Grid.Column="1"
            Style="{StaticResource button}"
            Content="HtmlDecode"
            Click="btn_HtmlDecode_Click"/>
        <TextBox x:Name="textBlock3"
            Grid.Row="2" Grid.Column="0"
            Style="{StaticResource textBox}"/>
        <Button x:Name="btn_UrlEncode"
            Grid.Row="2" Grid.Column="1"
            Style="{StaticResource button}"
            Content="UrlEncode"
            Click="btn_UrlEncode_Click"/>
        <TextBox x:Name="textBlock4"
            Grid.Row="3" Grid.Column="0"
            Style="{StaticResource textBox}"/>
        <Button x:Name="btn_UrlDecode"
            Grid.Row="3" Grid.Column="1"
            Style="{StaticResource button}"
            Content="UrlDecode"
```

```
            Click = "btn_UrlDecode_Click"/>
</Grid>
   C#：
public htmldom3()
{
        InitializeComponent();
        textBlock1.Text = "< div >这是一个 DIV </div>";
        textBlock2.Text = HttpUtility.HtmlEncode(textBlock1.Text);
        textBlock3.Text = "http://blog.csdn.net/dotfun";
        textBlock4.Text = HttpUtility.UrlEncode(textBlock3.Text);
}
private void btn_HtmlEncode_Click(object sender,RoutedEventArgs e)
{
        textBlock1.Text =
                    HttpUtility.HtmlEncode(textBlock1.Text);
}
private void btn_HtmlDecode_Click(object sender,RoutedEventArgs e)
{
        textBlock2.Text =
                    HttpUtility.HtmlDecode(textBlock2.Text);
}
private void btn_UrlEncode_Click(object sender, RoutedEventArgs e)
{
        textBlock3.Text =
                    HttpUtility.UrlEncode(textBlock3.Text);
}
private void btn_UrlDecode_Click(object sender, RoutedEventArgs e
{
        textBlock4.Text =
                    HttpUtility.UrlDecode(textBlock4.Text);
}
```

运行结果如图 5-5 所示。

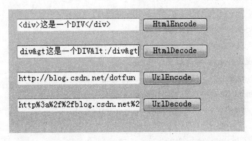

图 5-5　使用 HttpUtility 类的编码和解码

5.6.2　使用 Document.Cookies 读写 Cookie

浏览器的 Cookie 存放在客户端，可以用来存放用户的少量数据，并且可以在浏览器中分享 Cookie 信息。在 Silverlight 使用 HtmlPage.Document.Cookies 可以取得客户端全部的 Cookie 值。但是如果要取得单个的 Cookie 值，需要对获取的 Cookie 值进行遍历，示例如下。

```xaml
XAML:
<UserControl.Resources>
        <Style x:Key="textBox" TargetType="TextBox">
<Setter Property="Width" Value="200"/>
            <Setter Property="Height" Value="40"/>
            <Setter Property="FontSize" Value="20"/>
            <Setter Property="Margin" Value="10"/>
        </Style>
        <Style x:Key="textBlock" TargetType="TextBlock">
            <Setter Property="FontSize" Value="20"/>
            <Setter Property="Margin" Value="10"/>
        </Style>
        <Style x:Key="button" TargetType="Button">
            <Setter Property="Width" Value="120"/>
            <Setter Property="Height" Value="40"/>
            <Setter Property="FontSize" Value="15"/>
        </Style>
</UserControl.Resources>
<Grid x:Name="LayoutRoot" Background="GreenYellow">
        <Grid.RowDefinitions>
        <RowDefinition Height="80"></RowDefinition>
        <RowDefinition Height="80"></RowDefinition>
    </Grid.RowDefinitions>
    <Grid.ColumnDefinitions>
        <ColumnDefinition Width="780"></ColumnDefinition>
    </Grid.ColumnDefinitions>
    <StackPanel Grid.Row="0" Grid.Column="0"
        Orientation="Horizontal">
        <TextBlock Text="Name: "
                Style="{StaticResource textBlock}"/>
        <TextBox x:Name="txtSetName"
                Style="{StaticResource textBox}"/>
        <TextBlock Text="Value:"
                Style="{StaticResource textBlock}"/>
        <TextBox x:Name="txtSetValue"
                Style="{StaticResource textBox}"/>
        <Button x:Name="btnSetCookie"
                Style="{StaticResource button}"
                Content="SetCookie"
                Click="btnSetCookie_Click"/>
    </StackPanel>
    <StackPanel Grid.Row="1" Grid.Column="0"
        Orientation="Horizontal">
        <TextBlock Text="Name:"
                Style="{StaticResource textBlock}"/>
        <TextBox x:Name="txtGetName"
```

```xml
                        Style = "{StaticResource textBox}"/>
                <TextBlock Text = "Value:"
                        Style = "{StaticResource textBlock}"/>
                <TextBox x:Name = "txtGetValue"
                        Style = "{StaticResource textBox}"/>
                <Button x:Name = "btnGetCookie"
                        Style = "{StaticResource button}"
                        Content = "GetCookie"
                        Click = "btnGetCookie_Click"/>
</StackPanel>
</Grid>
```

C#:

```csharp
public htmldom4()
{
            InitializeComponent();
}
   private void btnGetCookie_Click(object sender,RoutedEventArgs e)
{
            //获取 Cookie 值
        txtGetValue.Text = CookieHelper.GetCookie(txtGetName.Text);
}
   private void btnSetCookie_Click(object sender,RoutedEventArgs e)
{
            //设置 Cookie 值
        CookieHelper.SetCookie(txtSetName.Text , txtSetValue.Text);
}
    //客户端 Cookie 读写类
public class CookieHelper{
        //根据 Key 和 Value 写客户端 Cookie
        public static void SetCookie(string key, string value)
        {
            DateTime expire = DateTime.UtcNow
                + TimeSpan.FromDays(30);
            string cookie = string.Format("{0} = {1};expires = {2}",
                key, value, expire.ToString("R"));
            HtmlPage.Document.SetProperty("cookie" , cookie);
        }
        //根据 Key 读客户端 Cookie
        public static string GetCookie(string key)
         {
            key += '=';
            //取出所有 Cookie
             string[] cookies =
                HtmlPage.Document.Cookies.Split(';'); //遍历 Cookie 值
            foreach (string cookie in cookies)
            {
                string cookieStr = cookie.Trim();
                //获取 Cookie 的 key 名称的位置
                if (cookieStr.StartsWith(key,
                    StringComparison.OrdinalIgnoreCase))
```

```
                    {
                        //分隔出 key 的值
                        string[] vals = cookieStr.Split('=');
                            if (vals.Length >= 2)
                            {
                                //返回值
                                return vals[1];
                            }
                        //如果没有找到则返回空白字符串
                        return string.Empty;
                    }
                }
                //如果没有 Cookie 则返回空白字符串
                return string.Empty;
            }
        }
```

运行结果如图 5-6 所示。

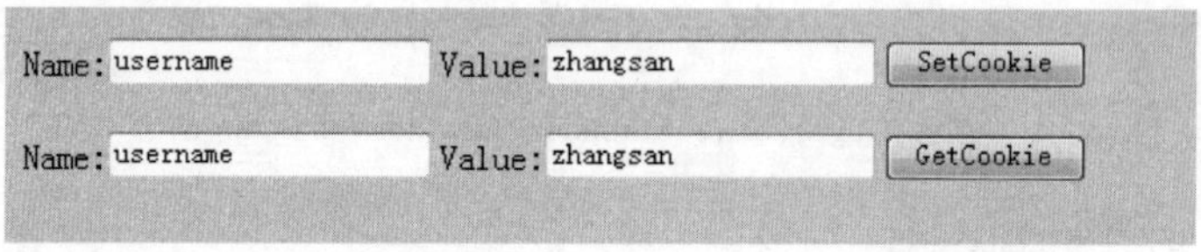

图 5-6 使用 Document.Cookies 读写 Cookie

CookieHelper 是我们自己编写的一个类，它提供了 SetCookie 和 GetCookie 的静态方法，分别根据 Key 读写客户端的 Cookie 值，在 Silverlight 中存储类似的内容除了使用传统的 Cookie 之外，还可以直接使用 Silverlight 推荐的独立储存特性 Isolated Storage。

5.6.3 使用 HtmlPage.Window 类

Silverlight 的 HtmlPage 中提供了 Window 类，Window 类封装了很多浏览器常用操作，例如消息框、输入框、对话框、页面导航等。下面通过两个示例——页面导航和消息提示介绍如何使用 HtmlPage.Window 类中常用的功能。

1. 页面导航示例

```
XAML:
<StackPanel Background = "GreenYellow"
            Orientation = "Horizontal">
    <TextBox x:Name = "tbUrl"
        Width = "400" Height = "40"
        Margin = "20"
        FontSize = "25"
        Text = "https://www.taobao.com"/>
    <Button x:Name = "btnNav"
        Width = "200" Height = "50"
        FontSize = "30"
```

```
                    Click = "btnNav_Click"
                    Content = "页面导航"/>
</StackPanel>
  C#:
private void btnNav_Click(object sender, RoutedEventArgs e)
{
            //根据输入值创建 URI 对象
        Uri uri = new Uri(tbUrl.Text, UriKind.RelativeOrAbsolute);
         //导航到 URI 地址
        HtmlPage.Window.Navigate(uri);
}
```

运行效果如图 5-7 和图 5-8 所示。

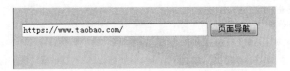

图 5-7　页面导航

图 5-8　导航页面

2．消息提示示例

```
XAML:
< Grid x:Name = "LayoutRoot" Background = "#C5FFA5">
        < TextBlock Padding = "30,30"
                  FontSize = "30"
                 Text = "在 Silverlight 中调用浏览器消息框"/>
        < StackPanel Orientation = "Horizontal" Margin = "50">
< Button x:Name = "button1" Click = "button1_Click"
            Content = "Alert" Margin = "20"
            Width = "180" Height = "50"
            FontSize = "25"
            Background = "Red"/>
         < Button x:Name = "button2" Click = "button2_Click"
            Content = "Confirm" Margin = "20"
            Width = "180" Height = "50"
            FontSize = "25"
```

```
                    Background = "Red"/>
            <Button x:Name = "button3" Click = "button3_Click"
                Content = "Prompt" Margin = "20"
                Width = "180" Height = "50"
                FontSize = "25"
                Background = "Red"/>
        </StackPanel>
</Grid>
```
C#:
```csharp
private void button1_Click(object sender, RoutedEventArgs e)
{
    HtmlPage.Window.Alert("这是使用 HtmlPage 调用的消息框!");

}
private void button2_Click(object sender, RoutedEventArgs e)
{
    if (HtmlPage.Window.Confirm("你确定吗?")
    {
    }
}
private void button3_Click(object sender, RoutedEventArgs e)
{
    string password = HtmlPage.Window.Prompt("请输入密码");
}
```

运行结果如图 5-9~图 5-11 所示。

图 5-9　Alert 消息提示

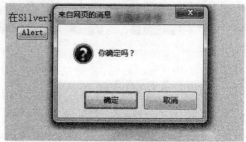

图 5-10　Confirm 消息提示

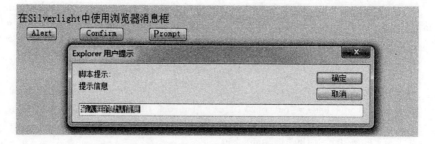

图 5-11　Prompt 消息提示

5.7 Silverlight 调用 JavaScript

JavaScript 是非常强大的客户端脚本语言，Silverlight 提供了与 JavaScript 进行交互的方法，用户不用关心它们之间的具体实现，看到的仅是程序运行的结果，所以可以说，Silverlight 某种意义上实现了 JavaScript 可以做的大部分事情。

Silverlight 提供了 Invoke 和 InvokeSelf 两种对 JavaScript 函数的访问接口，并且可以使用 HtmlPage.Document.CreateElement 方法在 Silverlight 程序中动态创建 JavaScript 脚本，下面通过两个示例介绍在 Silverlight 中创建和调用 JavaScript 的方法。

1. Silverlight 调用 JavaScript 示例之一

```
XAML:
<Grid x:Name = "LayoutRoot" Background = "#C5FFA5"
        Loaded = "LayoutRoot_Loaded">
<TextBlock Padding = "30,30"
            FontSize = "30"
            Text = "Silverlight 动态创建调用 JavaScript"/>
    <StackPanel Orientation = "Horizontal">
        <Button x:Name = "button1" Margin = "50"
            Click = "button1_Click"
            Content = "Invoke"
            Width = "200" Height = "100"
            FontSize = "25"/>
        <Button x:Name = "button2"
            Click = "button2_Click"
            Content = "InvokeSelf"
            Width = "200" Height = "100"
            FontSize = "25"/>
    </StackPanel>
</Grid>
C#:
private void button1_Click(object sender, RoutedEventArgs e)
{
        //使用 Invoke
        HtmlPage.Window.Invoke("calljs", "Invoke");
}
private void button2_Click(object sender, RoutedEventArgs e)
{
        //创建脚本对象
        ScriptObject calljs =
            (ScriptObject)HtmlPage.Window.GetProperty("calljs");
        //使用 InvokeSelf
        calljs.InvokeSelf("InvokeSelf");              }
private void LayoutRoot_Loaded(object sender, RoutedEventArgs e)
{
        //JavaScript 脚本
        string jsText = @"function calljs(msg){
```

```
            alert(msg);
        }";
        //创建脚本片段
        HtmlElement element =
                HtmlPage.Document.CreateElement("Script");
        element.SetAttribute("type", "text/javascript");
        element.SetProperty("text", jsText);
        //添加脚本到 Html 页面中
        HtmlPage.Document.Body.AppendChild(element);
}
```

运行结果如图 5-12 和图 5-13 所示。

图 5-12 Invoke 消息提示

图 5-13 InvokeSelf 消息提示

Invoke 和 InvokeSelf 都可以实现在 Silverlight 中调用 JavaScript 函数。Invoke 是 HtmlPage.Window 类的一个方法,InvokeSelf 使用 ScriptObject,通过 GetProperty 获取脚本对象,然后进行调用,CreateElement 用来创建 HTML 元素,这里用它产生一个 Script 片段,其结果与直接在 HTML 网页中声明< Script type="text/javascript">一样。

2．Silverlight 调用 JavaScript 示例之二

```
XAML:
< Grid x:Name = "LayoutRoot"
       Background = "#C5FFA5"
       Loaded = "LayoutRoot_Loaded">
    < TextBlock Padding = "30,30"
        FontSize = "30"
        Text = "使用 CreateInstance 获取 JavaScript 对象"/>
    < StackPanel Orientation = "Horizontal" Margin = "20">
        < TextBlock Text = "请输入文字:" Height = "50" FontSize = "25"/>
        < TextBox x:Name = "textBox1"
            Width = "300" Height = "50"
            FontSize = "20" Margin = "10" />
        < Button x:Name = "button1" Click = "button1_Click"
            Content = "显  示"
            Width = "150" Height = "50"
```

```
                FontSize = "25"
                Background = "Red"/>
        </StackPanel >
</Grid >
 C♯:
private void LayoutRoot_Loaded(object sender, RoutedEventArgs e)
    {
            //JavaScript 脚本
             string jsText = @"
            jsObject = function(msg)
            {
                    this.Msg = msg;
            }
            jsObject.prototype.Show = function()
            {
                    alert(this.Msg);
             }";
            //创建脚本对象
            HtmlElement element =
                HtmlPage.Document.CreateElement("Script");
            element.SetAttribute("type", "text/javascript");
            element.SetProperty("text", jsText);
            //添加 JavaScript 到 Html 页面
            HtmlPage.Document.Body.AppendChild(element);
        }
private void button1_Click(object sender, RoutedEventArgs e)
 {
            //使用 CreateInstance 获取 JavaScript 对象
            ScriptObject script =
                HtmlPage.Window.CreateInstance("jsObject",
                textBox1.Text);
            script.Invoke("Show");
 }
```

运行结果如图 5-14 所示。

图 5-14 使用 CreateInstance 获取 JavaScript 对象

这个示例与前面的示例的不同之处在于先使用了 HtmlPage.Window.CreateInstance 方法获取 ScriptObject 对象,然后再使用 Invoke 调用。因为这里的 jsObject 在 JavaScript 中是以一个对象变量来声明的,而不是一个单纯的 JavaScript 函数。

5.8 使用 JavaScript 调用 Silverlight

前面已经介绍了如何在 Silverlight 中创建和调用 JavaScript 对象函数,下面反过来介绍如何在 JavaScript 中调用 Silverlight 中的 C#方法。其实方法很简单,只要在 Silverlight 中注册一个 JavaScript 脚本对象,并将 C#方法暴露出来即可。下面介绍在 JavaScript 调用 Silverlight 的方法。

```
C#:
  public javascript5()
{
        InitializeComponent();
        //注册 JavaScript 的访问对象
        HtmlPage.RegisterScriptableObject("Builder", this);
}
    //定义 CreateRect 为脚本成员
    [ScriptableMember] //脚本化此方法
public void CreateRect(int width, int height)
 {
        //创建一个矩形对象
        Rectangle rect = new Rectangle();
}
```

运行结果如图 5-15 所示。

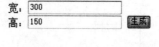

图 5-15 使用 JavaScript 调用 Silverlight

这个示例是使用 HTML 的文本框返回宽度和高度值。Silverlight 中的 CreateRect 方法接收到通过 JavaScript 传入的参数,并根据这个尺寸动态创建一个 Rectangle 对象。RegisterScriptableObject 用 CreateRect 方法给客户端脚本,注册 JavaScript 脚本对象。

总 结

本章介绍了 Silverlight 插件的常用参数,以及如何实现 Silverlight、JavaScript、HTML 三者交互。通过学习本章内容,有助于熟悉如何将 JavaScript 应用程序嵌入页面中,并实现 Silverlight 应用程序与 HTML 页面相互调用的接口方法。

所以,一个 Silverlight 应用程序在网页中并不是完全孤立的,它既可独当一面,也可以与其他网页内容集成和交互。

作 业

1. 简述 Silverlight 调用 JavaScript 需要经过几个步骤,分别是什么意思?
2. 新建一个 Silverlight 工程,包含用户名(文本框)、密码(文本框)、确定(按钮)和退后(按钮),单击"确定"按钮时,调用 JavaScript 验证文本框是否为空,如果为空则给出相应的提示。效果如图 5-16 所示。

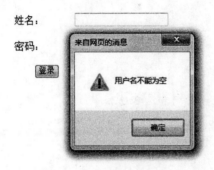

图 5-16 验证用户名和密码

第6章

数据访问与 Silverlight 高级应用实例

学习目标

- 掌握 Silverlight 中数据访问和远程通信
- 熟练使用 ServeFileDialog 和 OpenFileDialog
- 掌握 Silverlight 中的控件导航

6.1 数据访问与远程通信

Silverlight 提供了 WebClient 和 WebRequest 类,用来进行针对某一 URI 的简单访问,并可以把访问的结果返回给 Silverlight。WebClient 与 WebRequest 的作用类似,WebClient 可以实现最简单的远程通信。

6.1.1 WebClient

在 Silverlight 的早期版本中,WebClient 对应于 Downloader 对象,WebClient 支持"Write"和"Read"两种请求方向,分别是向 URI 进行写入请求和读取请求。下面介绍它们对应的事件和方法。

(1) Stream(流)读/写方法与事件如下。

OpenReadAsync 方法:根据打开的 URI 异步读取流。

OpenWriteAsync 方法:根据打开的 URI 异步写入流。

OpenReadCompleted 事件:读取结束。

OpenWriteCompleted 事件:写入结束。

(2) String(字符串)读/写方法与事件如下。

UploadStringAsync 方法:根据打开的 URL 上传字符串。

DownloadStringAsync 方法：根据打开的 URL 下载字符串。
UploadProgressChanged 事件：上传进度。
DownloadProgressChanged 事件：下载进度。
UploadStringCompleted 事件：上传字符串结束。
DownloadStringCompleted 事件：下载字符串结束。
示例代码如下：

```
XAML:
public webclientSample()
{
        InitializeComponent();                          //定义 URL 地址
            //创建 WebClient 对象
        string url = "http://localhost:1398/Sample.web/responseText.htm";
        WebClient webClient = new WebClient();          //定义异步请求地址
            //定义请求完成的事件处理
        webClient.DownloadStringAsync(
            new Uri(url, UriKind.RelativeOrAbsolute));
        webClient.DownloadStringCompleted
                += new DownloadStringCompletedEventHandler
                    (webClient_DownloadStringCompleted);
}
private void webClient_DownloadStringCompleted(object sender,
 DownloadStringCompletedEventArgs e)
{
        //显示返回值
        MessageBox.Show(e.Result.ToString());
}
```

这里声明了一个 URL 地址，为 WebClient 提供请求并返回字符串，WebClient 提供了 DownLoadStringAsync 和 OpenReadAsync 请求方法，当 WebClient 的请求完成后会进行回调，所以还需要定义完成后的事件处理程序 DownLoadStringCompleted 和 OpenReadCompleted。

从示例可以看到，回调事件中的 DownloadStringCompletedEventArgs 参数名为 e，它包含了返回的结果值 Result 属性，该属性返回这次请求的结果字符串。WebClient 还支持一个 AllowReadStreamBuffering 属性，它声明了通过 WebClient 打开指定的 URI 时，是否允许使用缓冲处理。

6.1.2　WebClient 与 XmlReader

XmlReader 支持对标准 XML 文件的读取。可以把一个 XML 文件作为一个字符串返回，然后将这个 Stream 交给 XmlReader 来读取。

下面就通过一个示例来介绍 XmlReader 对象来读取一个 XML 文件，同时使用 WebClient 去请求一个 XML 文件并返回 XML 文件的字符串结果。如果使用这种方式读取 XML 数据，那么在更新 XML 内容的同时，Silverlight 应用程序的内容会被更新，因为这个 XML 是通过 WebClient 来下载的，示例如下：

```
XAML:
public xmlReaderSample()
{
    InitializeComponent();
    LayoutRoot.Loaded += new RoutedEventHandler(LayoutRoot_Loaded);
}
private void LayoutRoot_Loaded(object sender, RoutedEventArgs e)
{
    tbRSS.Text = "http://localhost:1398/Sample.web/RSS-Source.xml";
}
private void btnGet_Click(object sender, RoutedEventArgs e)
{
    WebClient webClient = new WebClient();
    webClient.DownloadStringAsync(new Uri(tbRSS.Text,
        UriKind.RelativeOrAbsolute));
    webClient.DownloadStringCompleted
        += new DownloadStringCompletedEventHandler
            (webClient_DownloadStringCompleted);
}
private void webClient_DownloadStringCompleted(object sender,
 DownloadStringCompletedEventArgs e)
{
    //将结果字符串转换为字节进行读取
    List<Item> items = LoadRSS(Encoding.UTF8.GetBytes(e.Result));
    //使用泛型类 Item 作为 ListBox 的数据源
    lstRSS.ItemsSource = items;
}
private List<Item> LoadRSS(byte[] bytes)
{
    List<Item> items = new List<Item>();
    //将字节转换为 Stream 对象
    Stream stream = new MemoryStream(bytes);
    //加载 XML 并创建 XmlReader 对象
    XmlReader xmlReader = XmlReader.Create(stream);
    try
    {
        //找到 item 节点
        if (xmlReader.ReadToFollowing("item"))
        {
            //创建 Item 类
            Item item = new Item();
            //循环读取 item 节点中的内容
            while (xmlReader.Read())
            {
                if (xmlReader.IsStartElement("title"))
                {
                    item.Title = xmlReader
                        .ReadElementContentAsString().Trim();
                }else if(xmlReader.IsStartElement("link"))
```

```csharp
                    {
                        item.Link = xmlReader
                            .ReadElementContentAsString();
                    }else if(xmlReader.IsStartElement("author"))
                    {
                        item.Author = xmlReader
                            .ReadElementContentAsString();
                    }else if(xmlReader.IsStartElement("pubDate"))
                    {
                        item.PubDate = xmlReader
                            .ReadElementContentAsString();
                    }else if(xmlReader.IsStartElement(
                      "description"))
                    {
                        item.Description = xmlReader
                            .ReadElementContentAsString();
                    }else if(xmlReader.IsStartElement("item"))
                    {
                        items.Add(item);
                        item = new Item();
                    }
                }
                    items.Add(item);
            }
                return items;
        }catch{
                //显示错误信息
                HtmlPage.Window.Alert("RSS 获取错误!");
                return new List<Item>();
        }
}
private void lstRSS_SelectionChanged(object sender,
 SelectionChangedEventArgs e)
 {
     //显示一个 Item 的内容
     Item item = lstRSS.SelectedItem as Item;
     if (item != null)
     {
         tbTitle.Text = item.Title;
         tbLink.Text = item.Link;
         tbDate.Text = item.PubDate;
         tbDesc.Text = item.Description;
     }
 }
//Item 类
public class Item
{
        public string Title { get; set; }
        public string Link { get; set; }
```

```
            public string Author { get; set; }
            public string PubDate { get; set; }
            public string Description { get; set; }
        }
```

运行结果如图 6-1 所示。

图 6-1 运行结果

这是一个完整的示例，首先使用 WebClient 完成一个 XML 文件的读取，并将结果返回给一个 XmlReader。这个 XmlReader 负责处理 XML 文件的内容，将这个 XML 内容标题显示在 ListBox 控件里。代码中定义了一个 Item 类，这个类包括了 XML 节点的内容项，单击 ListBox 列表中的任何一项时，右边的区域会将这条内容分类显示出来，如图 6-2 所示。

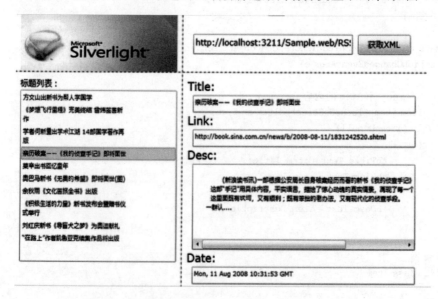

图 6-2 运行结果（单击 ListBox 列表中的任何一项后）

本示例中应用的 XmlReader 方法和作用如下。

(1) ReadToFollowing：定位到某个 XML 节点。
(2) Read：读取 XML 文件内容。
(3) IsStartElement：是否为指定名称的节点。
(4) ReadElementContentAsString：读取节点内容字符串。

6.1.3　WebRequest

Silverlight 中的 WebRequest 与 ASP.NET 的同名对象作用类似，它针对某一 URI 进行请求并返回值。这种请求就好比普通网页 form 的 submit，分为 POST 和 GET 两种方式，可以使用 Method 属性指定请求类型，使用 ContentType 属性指定请求内容的类型。WebRequest 的请求是异步的，请求过程是异步回调，通过 BeginGetResponse 开始，使用 AsyncCallback 指定回调方法。本示例中使用 Silverlight 的 WebRequest 对象请求一个普通的网页，并将网页中的内容返回 Silverlight 的 TextBox 控件中显示出来。示例代码如下。

```
XAML:
<Grid x:Name = "LayoutRoot" Background = "White">
    <TextBox x:Name = "tb" FontSize = "22" Background = "Green"
             HorizontalScrollBarVisibility = "Visible"
             VerticalScrollBarVisibility = "Visible"
             Foreground = "White" Text = "Loading..."/>
</Grid>
C#:
//定义异步委托方法
private delegate void DispatcherInvoke(string str);
public webrequestSample()
{
    InitializeComponent();
    //定义 URL 地址
    string url = "http://localhost:1398/Sample.web/responseText.htm";
    //创建 WebClient 对象
    WebRequest request = HttpWebRequest
            .Create(new Uri(url,UriKind.RelativeOrAbsolute));
    //开始获取响应并进行异步回调
    request.BeginGetResponse(new AsyncCallback(responseReady),
        request);
}
private void responseReady(AsyncResult ar)
{
        //返回异步回调结果对象
        WebRequest request = ar.AsyncState as WebRequest;
         //获取结果响应对象
        WebResponse response = request.EndGetResponse(ar);
        //定义返回流对象
        using (Stream stream = response.GetResponseStream())
        {
            //使用流读取对象
```

```
                    StreamReader reader = new StreamReader(stream);
                    // *** 直接读取将发生错误
                    //tbk.Text = "reader.ReadToEnd();
                    //使用 Dispatcher 异步执行委托方法
                    tb.Dispatcher.BeginInvoke
                            ((DispatcherInvoke)processResult,
                            reader.ReadToEnd());
                }
        }
        private void processResult(string result)
        {
                //显示返回字符串
                tb.Text = result;
        }
```

运行结果如图 6-3 所示。

图 6-3　使用 Silverlight 的 WebRequest 对象请求一个普通的网页并显示于文本框中

在使用 WebRequest 时：

（1）WebRequest 请求的 URL 必须是绝对地址（UriKind.Absolute）。

（2）请求的服务地址必须与 Silverlight 在同一域下，否则需要使用跨域策略，后面会介绍实现跨域方法。

（3）AsyncCallback 回调请求的线程和 Silverlight 的 UI 之间是非同步线程，所以用户不能在回调方法中操作界面中的对象，否则将出现"跨线程访问无效"的错误信息，因此示例中通过 Dispatcher 的异步线程请求方法 BeginInvoke 调用委托方法 DispatcherInvoke 实现在 TextBox 控件中显示返回的字符串内容。

6.1.4　实现跨域访问

跨域访问是指位于 A 站点（www.contoso.com）的内容试图通过 WebClient 或 WebRequest 的方式请求 B 站点（www.fabrikam.com）的任意 URI。

在 Silverlight 中对跨域访问有比较严格的限制，不能任意访问获取除了该 Silverlight 应用程序所在域以外的内容，如果想实现"跨域访问"，需要使用显式的跨域策略实现。

Silverlight 支持两种跨域文件,分别为 Silverlight 跨域策略(clientaccesspolicy.xml)和 Flash 跨域策略(crossdomain.xml)的子集,Silverlight 的跨域策略文件是比较简单的 XML 格式文件,如果想声明外域的 Silverlight 应用程序可以访问本域的所有服务。

clientaccesspolicy.xml 文件格式如下:

```
Clientaccesspolicy.xaml:
<?xml version="1.0" encoding="utf-8" ?><access-policy>
    <cross-domain-access>
      <policy>
       <allow-from http-request-headers="*">
          <domain uri="*"/>
       </allow-from>
       <grant-to>
          <resource path="/" include-subpaths="true"/>
       </grant-to>
      </policy>
    </cross-domain-access>
</access-policy>
Crossdomain.xml:
<?xml version="1.0" encoding="utf-8" ?>
<cross-domain-policy>
  <allow-access-from domain="*"/>
</cross-domain-policy>
```

clientaccesspolicy.xml 和 crossdomain.xml 文件必须部署在整个网站的根目录下才能生效。Silverlight 的请求首先会从整个网站的根目录中查找 clientaccesspolicy.xml 文件,如果没有发现就会查找 crossdomain.xml 文件。Domain 的值为"*"时,将服务公开为所有网站均可以访问,path 用来指定可以访问服务的路径,include-subpaths 声明是否包括 path 路径下的子路径的访问权限。

6.1.5 Silverlight 调用 WCF 服务

System.ServiceModel 命名空间赋予了 Silverlight 对各种 Web 服务的访问能力,其中包括 WebService 和 WCF,Silverlight 通过.NET 的 WebService 和 WCF 服务可以完成很多 Silverlight 自身功能以外的事情,由于 WebService 和 WCF 的操作方法类似,这里主要介绍 Silverlight 调用 WCF 的方法。

多数情况下,常在 Silverlight 中调用 WCF 返回某类数据集合或执行某种数据相关操作,这些集合的类型要求是 WCF 支持的可序列化的数据类型。下面将按步骤一步一步地介绍在 Web 站点中创建 WCF 服务并在 Silverlight 应用程序中调用它返回集合的过程。

(1) 在 Silverlight 的托管网站中创建一个名为 Service.SVC 的 WCF 服务,这时会在 Web 站点的 App-Code 目录中自动生成 Service.CS 类和 IService.CS 接口,以及 Service.SVC 文件。

(2) 在 Web 站点项目中添加一个名为 Log(日志)的类文件,这个类用于在 Silverlight 与 WCF 之间传递数据。

（3）在 Log 类中引用 System.Runtime.Serialization 命名空间，并添加类的属性成员，代码如下（这里为了简化内容只用了一个属性 Title，实际应用中可以添加任意多个属性成员）。

```
Log.cs: public class Log {
    //构造函数,带有两个参数
    public Log(string _title)
    {
        Title = _title;
    }
    //声明属性成员 Title
        [DataMember]
    public string Title {
    get; set;
    }
}
```

（4）接下来，为 Service.CS 和 IService.CS 添加接口方法和接口方法的实现，代码如下。

```
Iservice.cs: [ServiceContract] public interface IService {
    [OperationContract]Log[] GetLog(); }
Service.cs:
 //继承 IService 接口
public class Service : IService {
    //实现接口方法
    GetLog public Log[] GetLog(){
        Log[] log = {
            new Log("今天天气真不错,准备一起出去打球吧!"),
            new Log("微软的 Silverlight 技术真的很棒!我一定要学会使用它。"),
            new Log("这些内容是 WCF 返回给 Silverlight 的数据。")
        };
        return log;
    }
}
```

（5）上面已经完成了 WCF 服务的基本功能，在浏览器中输入 WCF 服务的名称就可以显示 WCF 服务信息，如图 6-4 所示。

这个 WCF 服务的作用就是返回一个 Log 类的数据集合。目前 Silverlight 支持的 WCF 通信类型是 basicHttpBinding，这就需要修改网站项目的 Web.config 文件 system.serviceModel 节点 binding 属性值为 basicHttpBinding，代码如下：

```
<services>
<service behaviorConfiguration = "ServiceBehavior" name = "Service">
    <endpoint address = "" binding = "basicHttpBinding"
        contract = "IService">
```

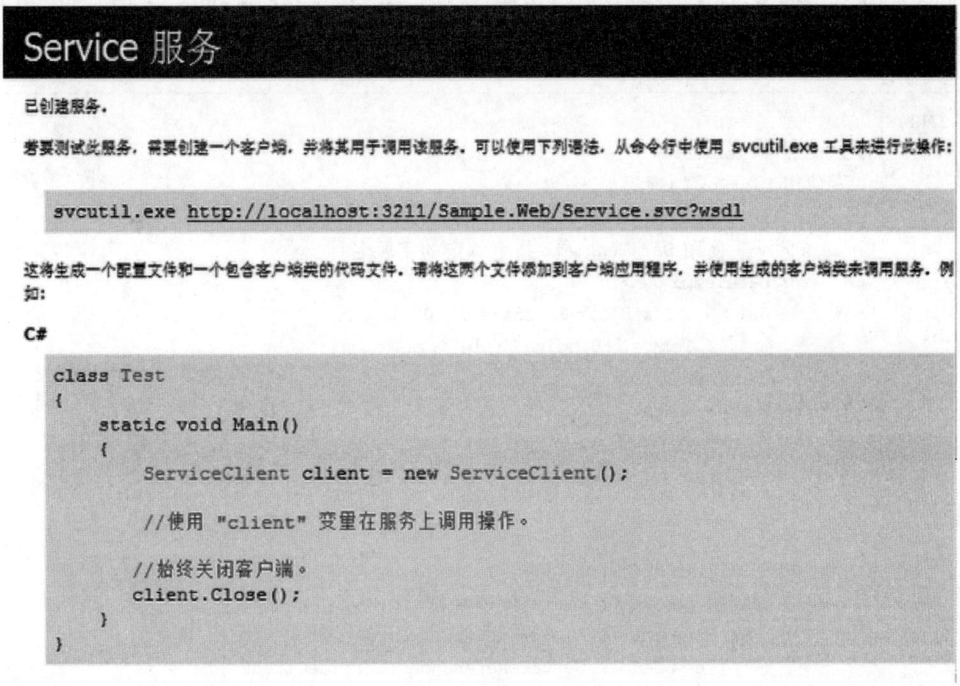

```
Service 服务

已创建服务.

若要测试此服务,需要创建一个客户端,并将其用于调用该服务。可以使用下列语法,从命令行中使用 svcutil.exe 工具来进行此操作:

    svcutil.exe http://localhost:3211/Sample.Web/Service.svc?wsdl

这将生成一个配置文件和一个包含客户端类的代码文件。请将这两个文件添加到客户端应用程序,并使用生成的客户端类来调用服务。例如:

C#

class Test
{
    static void Main()
    {
        ServiceClient client = new ServiceClient();

        //使用 "client" 变量在服务上调用操作。

        //始终关闭客户端。
        client.Close();
    }
}
```

图 6-4 在浏览器中输入 WCF 服务的名称就可以显示 WCF 服务信息

```
                < identity >
                    < dns value = "localhost"/>
                </identity >
        </endpoint >
        < endpoint address = "mex" binding = "mexHttpBinding"
            contract = "IMetadataExchange"/>
</service >
< service behaviorConfiguration = "DGServiceBehavior" name = "DGService">
        < endpoint address = "" binding = "basicHttpBinding"
          contract = "IDGService">
                < identity >
                    < dns value = "localhost"/>
                </identity >
        </endpoint >
        < endpoint address = "mex" binding = "mexHttpBinding"
            contract = "IMetadataExchange"/>
</service >
```

(6) 下面需要在 Silverlight 应用程序中引用这个 WCF 服务,右击 Silverlight 项目,从弹出的快捷菜单中选择"添加服务引用"命令。

(7) 在"添加服务引用"对话框中单击"发现"按钮,这时 Silverlight 会自动查找本地网站中所有可用服务并显示,这里显示出刚创建的名为 Service.SVC 的 WCF 服务,创建的引用名称默认为 ServiceReferencel,单击"确定"按钮,即完成了 Silverlight 应用对 WCF 服务的引用。

(8) 完成了 WCF 服务的引用后，开始为 Silverlight 程序制作展示界面，这里使用 ListBox 控件来显示数据列表，代码如下：

```
XAML:
<ListBox x:Name = "lstWCF">
    <!-- 使用 ListBox 的项模板 -->
    <ListBox.ItemTemplate>
        <!-- 使用 DataTemplate 指定绑定的字段名称 -->
        <DataTemplate>
            <TextBlock FontSize = "20"
            Text = "{Binding Title}"/>
        </DataTemplate>
    </ListBox.ItemTemplate>
</ListBox>
  C#:
public wcfSample()
{
    InitializeComponent();
    //创建 WCF 服务调用的代理对象
    ServiceReference1.ServiceClient proxy =
        new Sample.ServiceReference1.ServiceClient();
    //声明调用完成的事件处理
    proxy.GetLogCompleted +=
        new EventHandler<Sample.ServiceReference1
        .GetLogCompletedEventArgs>
        (proxy_GetLogCompleted);
    //开始异步调用
    proxy.GetLogAsync();
}
  void proxy_GetLogCompleted(object sender, Sample.ServiceReference1.GetLogCompletedEventArgs e)
    {
        //为 ListBox 控件绑定数据结果集合
        lstWCF.ItemsSource = e.Result;
    }
```

运行结果如图 6-5 所示。

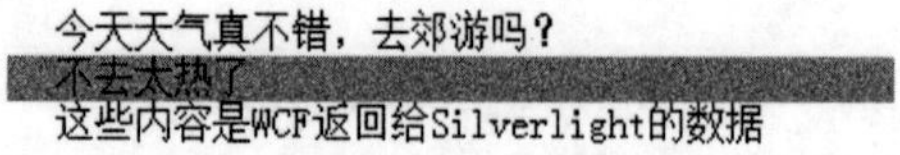

今天天气真不错，去郊游吗？
不去太热了
这些内容是WCF返回给Silverlight的数据

图 6-5 运行结果

在 Silverlight 调用 WCF 服务时，首先要在代理对象中创建一个服务，通过这个代理对象产生异步请求，在 XAML 界面中使用 ListBox 控件的 DataTemplate（数据模板），在模板中包含了一个 TextBlock 控件，它用来显示 Log 类 Title 的值，这样，一个完整的 WCF 服务调用就完成了，很简单。

这个示例仅展示了 Silverlight 调用 WCF 服务的基本方法。在实际应用中，WCF 可以做很多事情，例如操作文件、数据库等，这样使得 Silverlight 更加无所不能。

6.2 文件打开对话框与文件上传

在传统的网页上，用户一般是使用 HTML 中 File 标记实现对客户端文件的查看和选取，然后通过这个 File 元素 POST 到提供上传服务的网页中完成文件上传。

在 Silverlight 中可以使用文件对话框 OpenFileDialog 对象实现这一功能，该对象的 ShowDialog 方法可以弹出文件选择对话框，OpenFileDialog.File.OpenRead 方法用来读取选择的文件，使用前应当引用 System.iO 命名空间。

OpenFileDialog 仅是实现 Silverlight 客户端用户对本地的文件选取功能，在一个 Silverlight 应用程序示例中单纯完成文件的选取是没有实际意义的。下面通过一个示例演示 Silverlight 的文件对话框的使用方法，并通过 ASP.NET 的上传处理服务完成用户本地图片的上传功能，具体代码如下：

```
XAML:
<Grid x:Name = "LayoutRoot" Background = "White">
    <Grid.RowDefinitions>
        <RowDefinition Height = "280"/>
        <RowDefinition Height = " * "/>
    </Grid.RowDefinitions>
    <!-- 图片预览区域 -->
    <Border Grid.Row = "0" Grid.Column = "0" Width = "320" Height = "240"
        BorderBrush = "Black" BorderThickness = "2">
        <Image x:Name = "image" Stretch = "Fill"/>
    </Border>
    <!-- 功能区域 -->
    <StackPanel Orientation = "Horizontal" Width = "320"
        Grid.Row = "1" Grid.Column = "1">
        <Button x:Name = "btnPreview" FontSize = "23"
            Click = "btnPreview_Click"
            Content = "Preview"
            Width = "150" Height = "50"/>
        <Button x:Name = "btnUpload"
            FontSize = "23" Click = "btnUpload_Click"
            Content = "Upload" Width = "150" Height = "50"/>
    </StackPanel>
</Grid>
    C#:
public partial class OpenFileSample : UserControl
{
    private string filename = string.Empty;
    //创建文件对话框
    private OpenFileDialog openFile = new OpenFileDialog();
    public OpenFileSample()
    {
        InitializeComponent();
```

```csharp
    }
    void webclient_OpenWriteCompleted(object sender,
    OpenWriteCompletedEventArgs e)
    {
        //将图片数据流发送到服务器上
        Stream inputStream = e.UserState as Stream;
        Stream outputStream = e.Result;
        //缓冲区
        byte[] buffer = new byte[4096];
        int bytesRead = 0;
        while ((bytesRead = inputStream.Read(buffer, 0,
            buffer.Length)) > 0)
        {
            outputStream.Write(buffer, 0, bytesRead);
        }
            //关闭流
            outputStream.Close();
            inputStream.Close();
            MessageBox.Show("上传完成!");
    }
    private void btnUpload_Click(object sender, RoutedEventArgs e)
    {
        //判断用户是否已选择了图片
        if (openFile.File != null)
        {
            //图片文件名
            filename = "myPhoto.jpg";
            WebClient webclient = new WebClient();
            //指定上传处理的服务地址
            Uri uri = new Uri(
                http://localhost:1398/Sample.Web/
                UploadFile.ashx?filename=
                + filename, UriKind.Absolute);
            webclient.OpenWriteCompleted +=
                new OpenWriteCompletedEventHandler
                    (webclient_OpenWriteCompleted);
            //POST 图片文件到服务器 UploadFile.ashx
            webclient.OpenWriteAsync(uri, "POST",
                openFile.File.OpenRead());
        }else
        {
            MessageBox.Show("请先选取图片!");
        }
    }
    private void btnPreview_Click(object sender, RoutedEventArgs e)
    {
        //弹出文件对象框
        if (openFile.ShowDialog() == true)
        {
```

```
            BitmapImage bmp = new BitmapImage();
            //设置图源为用户选择的图片
            bmp.SetSource(openFile.File.OpenRead());
            image.Source = bmp;
        }
    }
}
```

程序运行后,当用户单击 Preview 按钮时,就会弹出文件选取窗口,如图 6-6 所示。

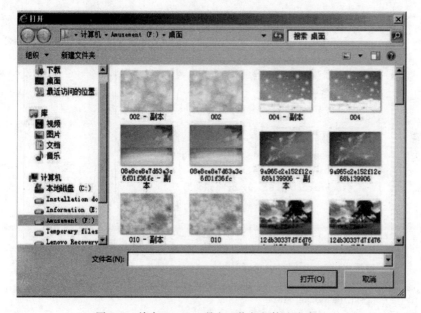

图 6-6　单击 Preview 按钮,弹出文件选取窗口

选择好一个图片文件,单击"打开"按钮,这时图片就会直接在界面上显示出来,如图 6-7 所示。

图 6-7　选择图片并打开

至此只是完成了 Silverlight 端的程序编写,接下来要添加服务器端的处理服务 UploadFile.ashx,其作用是将 POST 过来的文件通过 File 创建文件,然后使用 FileStream

方式写入服务器的目录里。UploadFile.ashx 代码如下：

```csharp
public void ProcessRequest (HttpContext context) {
    //获取上传的数据流
    Stream sr = context.Request.InputStream;
    int bytesRead = 0;
    byte[] buffer = new byte[4096];
    //取得文件名
    string filename = context.Request["filename"];
    try
    {
        //上传至服务器的 Upload 文件夹里
        string path =
            "D:\\Silverlight3\\Sample\\Sample.Web\\Upload\\";
        //将数据流写入服务器端
        using (FileStream fs =
        File.Create(path + filename, 4096))
        {
            while ((bytesRead =
                sr.Read(buffer, 0, buffer.Length)) > 0)
            {
                fs.Write(buffer, 0, bytesRead);
            }
        }
    }catch (Exception e)
    {
        //错误提示
        context.Response.ContentType = "text/plain";
        context.Response.Write
            ("上传失败，提示信息:" + e.Message);
    }finally
    {
        sr.Dispose();
    }
}
public bool IsReusable {
    get {
        return false;
    }
}
```

完成上面的服务端程序后，用户可以在 Silverlight 程序中先单击 Preview 按钮选择一个图片，然后单击 Upload 按钮，就会使用 WebClient 以 POST 的方式发送给服务端的处理页面 UploadFile.ashx。这时打开服务器端的 Upload 文件夹，就可以看到刚上传的图片文件，如图 6-8 所示。

第6章 数据访问与Silverlight高级应用实例 139

图 6-8 运行结果

6.3 使用保存文件对话框

SaveFileDialog 和 OpenFileDialog 都可以在客户端弹出文件对话框，其中 OpenFileDialog 是用来打开和选择一个客户端文件，而 SaveFileDialog 的功能正好相反，它用来将 Silverlight 应用程序中的内容保存到用户本地的保存文件对话框中。

当需要为 Silverlight 应用程序用户提供一些内容的保存功能时，SaveFileDialog 就有了用武之地，通过它可以让用户自行选择文件的保存位置，使用前请引用 System.IO 命名空间，示例如下：

```
XAML:
<Grid x:Name = "LayoutRoot" Background = "White">
    <StackPanel Orientation = "Vertical" Margin = "5">
        <TextBlock Height = "40" FontSize = "20"
                Text = "SaveFileDialog"/>
        <TextBox x:Name = "tbContent"
                Margin = "5" FontSize = "20"
                Text = "请在这里输入文字并点击保存!"
                Height = "200"
                AcceptsReturn = "True"/>
        <Button x:Name = "btnSave"
                Content = "SaveFile"
                Click = "btnSave_Click"
                Width = "120" Height = "40"
                FontSize = "18"/>
    </StackPanel>
</Grid>
```

运行结果如图 6-9 所示。

图 6-9　使用保存文件对话框

当输入文字并单击 SaveFile 按钮时，会调用 SaveFileDialog 的 ShowDialog 方法，弹出一个"另存为"对话框，输入文件名后单击"保存"按钮后，就可以在用户的本地文件夹中找到 Silverlight 保存的文件了，打开后可以看到刚才输入的文本内容，如图 6-10 所示。

图 6-10　运行结果

上面的示例实现了将用户输入的文件内容保存成一个文本文件到用户本地，其实 SaveFileDialog 不但可以保存 TXT 文件，还可以设置 Filter 为其他任何类型的文件。例如使用 SaveFileDialog 保存一个 HTML 网页，那么设置 Filter 的值是.html 或.htm 文件后缀名即可。HTML 文件可以包含文本、图片，甚至另一个 Silverlight 的 Object 等，开发者可以尽可能地发挥想象力，创意的空间是永远没有止境的。

6.4　启用 Silverlight 应用程序库缓存

应用程序库缓存是一种新的 Silverlight 应用程序打包与下载的机制，一个 Silverlight 应用程序的功能是基于不同类别的 DLL 库文件产生的，应用程序库缓存允许开发者仅将这个 Silverlight 应用程序的核心 DLL 打包，而其他部分可以根据需要由 Silverlight 自动下载。

XAP 文件是 Silverlight 应用程序发布和部署的一种形式,它包含 Silverlight 应用程序编译后的文件压缩包。当用户在浏览器上访问一个包含 Silverlight 应用程序的页面时,这个 XAP 文件将首先被下载到浏览器的缓存文件内,然后再由 Silverlight 插件运行并呈现出来。XAP 包的大小影响 Silverlight 应用程序的打开速度,只要减小 XAP 的大小,就会提高 Silverlight 应用程序的下载和打开的速度。

实现程序集缓存是很简单的,方法是在 Visual Studio 的解决方案资源管理器中选择一个 Silverlight 项目,然后查看项目属性,可以看到启用应用程序缓存功能的复选框,如图 6-11 所示。

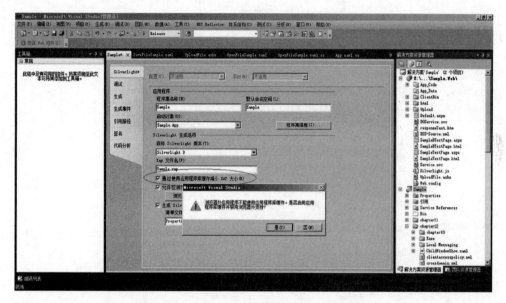

图 6-11　启用应用程序缓存功能的复选框

可以发现启用后的 XAP 文件中没有 Silverlight 项目中使用到的 DLL 库文件,它们都被打包放在 XAP 的同级文件夹里,Silverlight 应用程序在运行时会根据需求去下载相应的库文件,对于初始化简单的 Silverlight 应用程序而言,这样做可以使 Silverlight 程序很快地下载并显示,同时实现了 Silverlight 库文件的按需下载。

6.5　应用控件截图功能

Silverlight 3 的 System.Windows.Media.Imaging 命名空间新增了一个 WriteableBitmap,即可写入位图对象。WriteableBitmap 对 Silverlight 的 UIElement 派生对象提供 Render 方法,这样,就可以通过 WriteableBitmap 对各种 Silverlight 的 XAML 元素的外观进行截图,并且可以设置截取的尺寸,以及通过 Transform 对象为截图添加各种变形效果,这使得 WriteableBitmap 的作用更为灵活。

下面将演示如何使用 WriteableBitmap 对视频控件进行截图,并将截图的结果保存到 WrapPanel 控件中,这个示例用到的 WrapPanel 控件在 Silverlight 控件章节有相关的介绍。

WriteableBitmap 的示例如下：

```
XAML:
< StackPanel x:Name = "LayoutRoot" Background = "White">
    < controls:WrapPanel x:Name = "wrap"  Orientation = "Horizontal" />
        < Border HorizontalAlignment = "Center" CornerRadius = "5"
            BorderBrush = "Black" BorderThickness = "5">
            < MediaElement x:Name = "media" Stretch = "None"
                    Source = "/Video/SilverlightIntro.wmv"

                    MouseLeftButtonDown =
                        "media_MouseLeftButtonDown"/>
        </Border >
</StackPanel >
    C#:
//创建可写入位图对象,尺寸根据 MediaElement 的实际大小
    WriteableBitmap wBitmap
        = new WriteableBitmap(media, new MatrixTransform());
    //创建一个图像
Image img = new Image();
img.Width = 150;
img.Margin = new Thickness(10);
//将 wBitmap 作为图像源
img.Source = wBitmap;
//将图像添加到 WrapPanel 控件
wrap.Children.Add(img);
```

运行结果如图 6-12 所示。

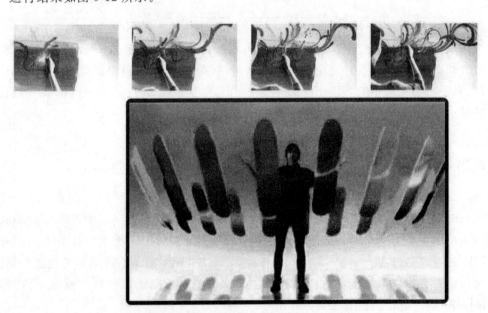

图 6-12　应用控件截图功能

WriteableBitmap 所生成的截图都依次保存在 WrapPanel 控件中,每当用户单击一次 MediaElement 对象,就会产生一幅当前视频画面的图像。

WriteableBitmap 还支持 Lock(锁定)和 UnLock(解除锁定)方法。当 WriteableBitmap 处于锁定状态时,不能通过 Render 方法更新 WriteableBitmap 对象,直到 WriteableBitmap 执行解除锁定方法为止。

6.6 Silverlight 3 Easing 动画集合

前面已经介绍了 Silverlight 动画的基本类型以及在 Silverlight 中创建和管理动画元素的方法。对于"动画"这个概念而言,除了用来实现动画效果本身的程序代码以外,更重要的是动画的创意与动画设计理念,这里面其实包含很多知识。

如果要实现自由落体运动、加速运动之类的动画效果,就涉及相关数学方面的诸多知识,很显然,这些并不是程序开发人员最擅长的方向,设计师也可能无从下手。针对这种相对某种固定模式的运动动画效果,在 Silverlight 3 中可以考虑使用 Easing Functions 模拟它,Easing Functions 可以称为简单动画函数集合,或开放动画集合,因为它提供了丰富的动画效果。

这些动画效果是从事先定义好的数字公式转化而来的,Ease 动画类的派生关系如下。

```
System.Object
   System.Windows.DependencyObject
      System.Windows.Media.Animation.EasingFuncionBase
         System.windows.Media.Animation.BackEase
         System.Windows.Media.Animation.BounceEase
         System.WindOWS.Media.Animation.CircleEase
         System.Windows.Media.Animation.CubiCEase
         System.Windows.Media.Animation.ElasticEase
         System.Windows.Media.Animation.ExponentialEase
         SyStem.WindOWS.Media.Animation.PowerEase
         System.Windows.Media.Animation.QuadraticEase
         System.Windows.Media.Animation.QuarticEase
         System.Windows.Media.Animation.QuinticEase
         System.Windows.Media.Animation.SineEase
```

所有 Ease 动画都是通过 EasingFunctionBase 类派生的。上面的每个派生类都代表一个 Ease 动画效果,所以每个 Ease 动画产生的变化是各不相同的,但是它们均支持 EasingMode 属性,这个属性可以改变动画运动的表现形式,它提供 3 种模式值。

(1) EaseIn:Easing 动画将数学公式的运算结果呈现出来。

(2) EaseOut:Easing 动画将数学公式的运算结果以完全相反的结果呈现出来,作用与 EaseIn 相反。

(3) EaseInOut:它将 EaseIn 和 EaseOut 两个部分效果联合起来成为一个动画,在整个动画运动的前半部分会产生 EaseIn 效果,后半部分则是 EaseOut 动画。

使用 Easing Functions 时，只需要在 Silverlight 动画元素（例如 DoubleAnimation）上添加并设置相应的属性（例如 BackEase）即可。

根据 System.Windows.Media.Animation.EasingFunctionBase 类的派生类，可以得到以下预设的 Ease 动画。

（1）BackEase（反向动画）

在产生正向动画前会进行一定幅度的反方向运动，Amplitude 属性代表振幅值。

（2）BounceEase（弹性动画）

创建弹起的效果，Bounces 属性代表活跃强度，数值越大强度越大，Bounciness 属性代表活跃高度，数值越大高度越小。

（3）CircleEase（圆周动画）

通过圆周函数创建加速或减速的动画。

（4）CubicEase（立方体动画）

创建加速或减速使用立方体公式的动画 $f(t)=t3$。

（5）ExponentiaEase（指数动画）

使用一个指数的公式创建加速或者减速的动画，Exponent 属性可以产生速度上的变化。

（6）PowerEase（动力动画）

使用 P＝Power 属性，即使用公式 $F(t)=tp$ 创建加速或者减速的动画，Power 属性可以控制动力值，值越大动力越强。

（7）QuadraticEase（二次方程动画）

使用公式 $f(t)=t2$ 创建加速或减速的动画。

（8）QuarticEase（四次方程动画）

使用公式 $f(t)=t4$ 创建加速或减速的动画。

（9）QuinticEase（次的方程动画）

（10）SineEase（正弦动画）

使用正弦公式创建加速或减速的动画。

前面描述的是数学的运动曲线，这些曲线图形并非是产生运动的轨迹。

6.7　使用墨迹控件 InkPresenter

InkPresenter 称为墨迹控件，继承于 Canvas 类。事实上它并不是一个单纯的控件，而应该把它看作 Silverlight 中的一种用户输入接口，因为它可以支持多种输入设备，并可以应用到不同的方面。

InkPresenter 的输入设备有手写笔输入、触摸板输入和鼠标输入。墨迹控件在 Web 应用程序中有很多用武之地，例如，可以使用墨迹为一些 Silverlight 中的文本内容添加批注，为博客或论坛添加个性签名，为联机游戏的输入交互开创新的应用局面……这一切实现起来非常简单，只需要在应用程序适当的位置添加 InkPresenter 控件，并保存 InkPresenter 的 Strokes 对象即可。

下面通过一个标准示例介绍如何使用墨迹控件 InkPresenter，代码如下：

```
XAML:
<InkPresenter x:Name = "IP"
              Width = "700" Height = "500"
              MouseMove = "IP_MouseMove"
              LostMouseCapture = "IP_LostMouseCapture"
              MouseLeftButtonDown = "IP_MouseLeftButtonDown">
    <InkPresenter.Background>
        <!--墨迹画笔的背景图片-->
        <ImageBrush ImageSource = "/../../../Images/SilverlightBlack.jpg"
            Stretch = "Fill"/>
    </InkPresenter.Background>
</InkPresenter>
```

```csharp
C#:
//定义一个画笔对象
private Stroke myStroke;
private void IP_MouseLeftButtonDown(object sender,
    MouseButtonEventArgs e)
{
    //墨迹画笔捕获鼠标
    IP.CaptureMouse();
    //创建画笔坐标集合 SPC
    StylusPointCollection SPC = new StylusPointCollection();
    //将当前 IP 对象的坐标集合添加到 SPC
    SPC.Add(e.StylusDevice.GetStylusPoints(IP));
    //将坐标集合添加到画笔对象
    myStroke = new Stroke(SPC);
    //墨迹的尺寸
    myStroke.DrawingAttributes.Width = 10;
    myStroke.DrawingAttributes.Height = 10;
    //墨迹的颜色
    myStroke.DrawingAttributes.Color = Colors.White;
    //将画笔对象添加到 IP 的画笔集合中
    IP.Strokes.Add(myStroke);
}
private void IP_MouseMove(object sender, MouseEventArgs e)
{
    if (myStroke != null)
    {
        /*当鼠标移动时,如果画笔不为空值就继续添加
        新的坐标点到 StylusPoints 集合*/
        myStroke.StylusPoints.Add(e.StylusDevice.GetStylusPoints(IP));
    }
}
private void IP_LostMouseCapture(object sender, MouseEventArgs e)
{
    //释放鼠标后将画笔置为空值
    myStroke = null;
}
```

示例的运行结果如图 6-13 所示。

图 6-13　使用墨迹控件 InkPresenter 手绘结果

DrawingAttributes 是 Stroke 的属性集合，包括对画笔样式的定义，例如宽度、高度、颜色，Strokes 负责收集多个 Stroke 对象，而一个 Stroke 对象是通过 StylusPointCollection 的坐标点集合生成的，各种输入设备的触点是通过 InkPresenter 事件中的 GetStylusPoints 方法进行收集的，可以根据代码中的注释了解程序属性的定义和作用。

以上仅实现了基本的墨迹输入功能，读者可以在以上程序的基础上进一步完善，例如添加背景画板的替换，对画笔的尺寸、画笔的颜色的制定等功能。

6.8　使用 Silverlight 控件导航

在 Silverlight 3 之前，当需要在一个 Web 应用程序站点中全部运用 Silverlight 技术呈现网站内容时，不得不使用 XAML 局部内容的更新满足这种需求。但使用这种方式存在一个令人很头痛的问题，就是当用户刷新这个浏览器的地址链接时，这个用户看到的整个 Silverlight 应用程序站点的内容又被重新加载，这样很可能会使用户丢失当前查看位置的数据。

举例说明，当用户在浏览一个某网站中的某个二级新闻页面时，这个页面中的新闻内容一般是"静态的"，就是说这些信息资料在用户没有刷新浏览器之前一般是不能够被更新的，除非使用 Timer 定时去更新它们。如果用户正在阅读内容时不小心刷新了浏览器链接地址，那么这个用户只能重复先前的操作才能再次到达用户关心的内容部分，因为一个 Silverlight 应用程序在浏览器中的地址是始终不变的。

值得庆幸的是，以上问题在 Silverlight 3 中得到了解决，因为 Silverlight 3 引入了称为 Navigation 的控件，即 Silverlight 应用程序导航控件。基本上 Navigation 可以帮助完成两件事：

（1）可以在不刷新整个页面的前提下，改变页面浏览器链接地址。

（2）当用户刷新浏览器时会根据 Silverlight 事先设置好的链接地址呈现页面内容。

这两点正是我们需要解决的问题，并且它们都是基于一个网页完成的，而不是在浏览器链接中改变页面 XAML。

当安装了基于 Silverlight 3 版本的 Silverlight Tools 后,可以在 Visual Studio 2008 的 Silverlight 项目模板找到 Silverlight Navigation Application(Silverlight 应用程序导航项目)。新建一个名为 NavSample 的 Silverlight 应用程序导航项目,单击"确定"后在解决方案资源管理器中可以看到以下项目结构,如图 6-14 所示。

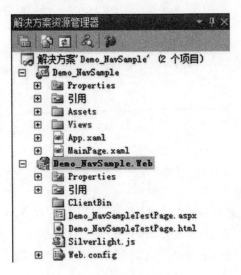

图 6-14　项目结构

这个示例给我们提供了学习 Navigation 的途径,可以直接运行它了解 Navigation 的作用,如图 6-15 所示。

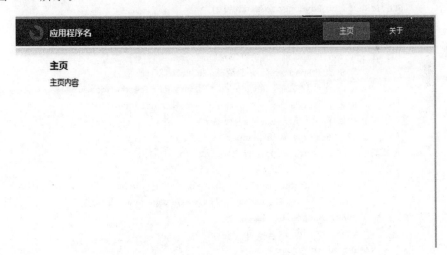

图 6-15　运行 Navigation 项目

这里承载 Silverlight 应用程序的是一个 NavSampleTestPage.aspx 的 ASP.NET 页面,当然也可以使用其他类型的网页来托管它。

默认情况下,应用程序具有两个导航链接按钮:Home(主页)和 About(关于)。一般来说"主页、关于"是网站的基本要素,当单击这个 Silverlight 程序中的 Home 或 About 链接按钮时,可以观察到整个浏览器标题和链接地址都发生了改变,如图 6-16 所示。

图 6-16　单击 About 链接按钮

这是 Silverlight 3 提供的导航应用程序项目模板，包含 MainPage.xaml、HomePage.xaml、ErrorWindow.xaml、AboutPage.xaml 和 App.xaml 页面。下面分别看一下这几个页面都有些什么秘密。首先看一下核心的页面 MainPage.xaml，这个页面是一个 UserControl 类型的 XAML 文件。

```xaml
XAML:
<Grid x:Name="LayoutRoot" Style="
    {StaticResource LayoutRootGridStyle}">
    <Border x:Name="ContentBorder" Style=
        "{StaticResource ContentBorderStyle}">
        <navigation:Frame x:Name="ContentFrame" Style=
            "{StaticResource ContentFrameStyle}"
            Source="/Home" Navigated="ContentFrame_Navigated"
            NavigationFailed="ContentFrame_NavigationFailed">
            <navigation:Frame.UriMapper>
                <uriMapper:UriMapper>
                    <uriMapper:UriMapping Uri=""
                        MappedUri="/Views/Home.xaml"/>
                    <uriMapper:UriMapping Uri="/{pageName}"
                        MappedUri="/Views/{pageName}.xaml"/>
                </uriMapper:UriMapper>
            </navigation:Frame.UriMapper>
        </navigation:Frame>
    </Border>
    <Grid x:Name="NavigationGrid" Style=
        "{StaticResource NavigationGridStyle}">
        <Border x:Name="BrandingBorder" Style=
            "{StaticResource BrandingBorderStyle}">
            <StackPanel x:Name="BrandingStackPanel" Style=
                "{StaticResource BrandingStackPanelStyle}">
                <ContentControl Style="{StaticResource LogoIcon}"/>
```

```xml
                    <TextBlock x:Name = "ApplicationNameTextBlock"
                        Style = "{StaticResource ApplicationNameStyle}"
                        Text = "应用程序名"/>
                </StackPanel>
            </Border>
            <Border x:Name = "LinksBorder"
                Style = "{StaticResource LinksBorderStyle}">
                <StackPanel x:Name = "LinksStackPanel" Style =
                    "{StaticResource LinksStackPanelStyle}">
                    <HyperlinkButton x:Name = "Link1" Style =
                        "{StaticResource LinkStyle}"
                        NavigateUri = "/Home" TargetName =
                        "ContentFrame" Content = "主页"/>
                        <Rectangle x:Name = "Divider1"
                        Style = "{StaticResource DividerStyle}"/>
                        <HyperlinkButton x:Name = "Link2"
                        Style = "{StaticResource LinkStyle}"
                        NavigateUri = "/About"
                        TargetName = "ContentFrame" Content = "关于"/>
                </StackPanel>
            </Border>
        </Grid>
    </Grid>
```

这个页面里面使用了大量的 Style 定义各个 XAML 控件的元素。可以看出，这是一种类似传统网站的开发流程，传统网站通过 CSS 样式统一整个网站样式，而在 Silverlight 中对应的是 Style。

 提 示

当然也可以把样式写在单独的 XAML 文件里。

在 MainPage.xaml 中还可以找到导航按钮和页面导航控件 Navigation。navigation：Frame 是指导航的类型，navigation 里包含 Frame 和 Page，Page 是指可以被 Frame 包含的一块 XAML 内容，这个 XAML 内容就是 navigation：Page，而导航控件就是 navigation：Frame。可以把它看作 Silverlight 中的 iframe 元素，HTML 中的 iframe 可以包含其他 HTML 页面，而 Silverlight 中 navigation：Frame 可以包含任何 navigation：Page 的 XAML 页面，这样是不是就容易理解了呢。可以注意到 navigation：Frame 的 Source 属性，它就是指定包含的 Page 页面。

接下来看一下 navigation 是如何实现导航功能的，导航按钮具有 Tag 属性，属性上定义了 HomePage 和 AboutPage 的 XAML 文件链接，当用户单击页面时页面 URL 的"♯"号的后面部分就会变成对应的 Tag 值，而导航控件会自动获取这个值，并加载相应的 navigation：Page，所以不需要编写具体代码，Navigation 控件把这一切都做好了，看一下 MainPage.xaml.CS 的 C♯代码。

```
MainPage.xmal.cs:
//在 Frame 导航之后,请确保选中表示当前页的
HyperlinkButton private void ContentFrame_Navigated(object sender,
NavigationEventArgs e)
{
    foreach (UIElement child in LinksStackPanel.Children)
    {
        HyperlinkButton hb = child as HyperlinkButton;
        if (hb != null && hb.NavigateUri != null)
        {
            if(hb.NavigateUri.ToString().Equals(e.Uri.ToString()))
            {
                VisualStateManager.GoToState(hb,"ActiveLink",true);
            }else
            {
                VisualStateManager.GoToState(hb,"InactiveLink",true);
            }
        }
    }
}
        //如果导航过程中出现错误,则显示错误窗口
private void ContentFrame_NavigationFailed(object sender,
 NavigationFailedEventArgs e)
{
        e.Handled = true;
        ChildWindow errorWin = new ErrorWindow(e.Uri);
        errorWin.Show();
}
```

上面 C#代码的 NavButton Click 事件处理程序的代码简单,只是把 Button 的 Tag 值取到,最关键的就是 Frame.Navigate 方法,它使 Navigation 控件获知需要加载哪个 Page 的方法,通过这个 Navigate 方法就可以实现 Silverlight 的页面导航,下面看一下 Home 和 About 的页面代码。

```
Home.xaml:
< Grid x:Name = "LayoutRoot">
    < ScrollViewer x:Name = "PageScrollViewer"
        Style = "{StaticResource PageScrollViewerStyle}">
        < StackPanel x:Name = "ContentStackPanel">
            < TextBlock x:Name = "HeaderText"
                Style = "{StaticResource HeaderTextStyle}" Text = "主页"/>
            < TextBlock x:Name = "ContentText" Style =
            "{StaticResource ContentTextStyle}" Text = "主页内容"/>
        </StackPanel>
    </ScrollViewer>
</Grid>
```

```
About.xaml:
< Grid x:Name = "LayoutRoot">
        < ScrollViewer x:Name = "PageScrollViewer"
         Style = "{StaticResource PageScrollViewerStyle}">
            < StackPanel x:Name = "ContentStackPanel">
                < TextBlock x:Name = "HeaderText"
                Style = "{StaticResource HeaderTextStyle}"
                Text = "关于"/>
                < TextBlock x:Name = "ContentText"
                Style = "{StaticResource ContentTextStyle}"
                Text = "关于页内容"/>
            </StackPanel>
        </ScrollViewer>
</Grid>
```

上面两个 XAML 页面的代码非常简单,只有简单的文本,可以修改它们然后重新编译程序改变主页和"关于"页的内容,也可以添加更多的导航链接和对应的页面。

提 示

添加的页面应该是一个 Silverlight Page,而不是一个 silverlight UserControl。

有 Web 开发经验的读者可能会从页面的链接上发现,Silverlight 的 Navigation 控件就是通过改变一个 URL 的"锚点链接"实现无刷新的 URL 更新。这样,在页面的 URL 更新的同时,navigation:Frame 就会动态地加载对应的页面。这类似 HTML 中的框架。

总 结

本章介绍了 Silverlight 的数据访问和通信技术,读完本章后相信读者对 Silverlight 的了解会更进一步。其实就 Silverlight 技术本身而言,它仅是一个客户端浏览器的插件程序,但是它能够通过多种手段实现对服务器的各种服务进行访问和调用,并将服务端的结果返回到 Silverlight 应用程序中展示出来,适当使用 Silverlight 的这些通信特性可以使开发者有更大的发挥空间,也使 Silverlight 应用程序的功能变得更加丰富。

作 业

1. 结合上机练习完成对用户权限的设置。
2. 管理员对员工权限的修改和删除。
3. 用 Blend 设计界面使其美观,局部呈现 3D 效果,例如登录界面有投影效果。

上机部分

土地改革

上机1

Silverlight 概述

上机任务

任务1　新建一个 Silverlight 程序
任务2　初识 Blend
任务3　使用 Blend 工具制作简单动画

第1阶段　指导

指导1　新建一个 Silverlight 程序

涉及重要知识点：
➢ 使用 Visual Studio 2010 新建 Silverlight 应用程序
➢ Silverlight 布局

问题

通过第1章课程的学习，大家对什么是 Silverlight、Silverlight 优势以及为什么要选择 Silverlight 都有一定的了解，下面来创建第一个 Silverlight 程序，如图上机 1-1 所示。

图上机 1-1　创建第一个 Silverlight 程序

解决方案

（1）打开 Visual Studio 201，创建一个"空解决方案项目"；
（2）在 F：\\Web 目录创建一个网页文件，命名为 Demo1；
（3）选择"在新网站中承载 Silverlight 程序"。

指导 2 初识 Blend

涉及重要知识点：
➢ 了解 Blend 基本布局

问题

在以前的 WPF 中大家可能接触过 Blend，那么现在再来看看 Blend 中的界面布局是怎么样的，以及如何使用 Blend 创建 Silverlight 程序。

解决方案

Silverlight 开发环境搭建成功后，将通过实例介绍 Blend 入门操作，首先使用 Blend 按照以下步骤创建一个新的 Silverlight 项目：

单击左上角的 File→New Project，这时会弹出一个新建项目窗口，如图上机 1-2 所示。

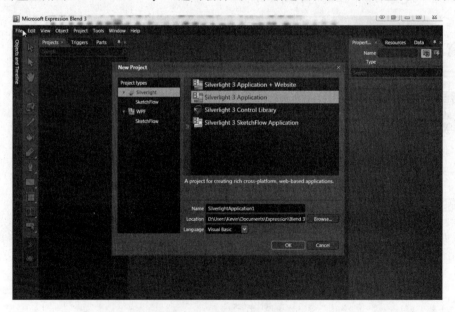

图上机 1-2 创建 Blend 程序

从图上机 1-2 可以看到 Blend 支持创建两个类型的项目，一个是 Silverlight，另一个是 WPF。本书主要讨论 Silverlight，所以，选择 Project Type→Silverlight，选中 Silverlight 类型后，在弹出窗口右边会出现 4 个项目模板，具体如下。

➢ Silverlight 3 Application＋Website

该选项是"创建 Silverlight 3 客户端应用其中包含 Website 项目"。选择该选项后，

Blend 会自动创建 Web 项目并在同一个解决方案下,编译后运行,会在 Web 项目中产生测试页面,在客户端显示 Silverlight 客户端。

➤ Silverlight 3 Application

该选项是"创建 Silverlight 3 客户端应用"。选择该选项后,Blend 仅创建 Silverlight 客户端,编译后,自动生成一个测试页面。

➤ Silverlight 3 Control Library

该选项是"创建 Silverlight 控件类库"。选择该选项后,Blend 会创建 Silverlight 空白类库,主要用于创建 Silverlight 自定义控件。

➤ Silverlight 3 SketchFlow Application

该选项是"创建 Silverlight 3 SketchFlow 应用"。选择该选项后,Blend 会创建 Silverlight 3 SketchFlow 应用。

在项目模板窗口下,是创建项目的名称、项目路径和项目后台语言支持。

创建一个 Silverlight 3 Application+Website+C♯完整项目,方便以后解释项目细节。如图上机 1-3 所示。

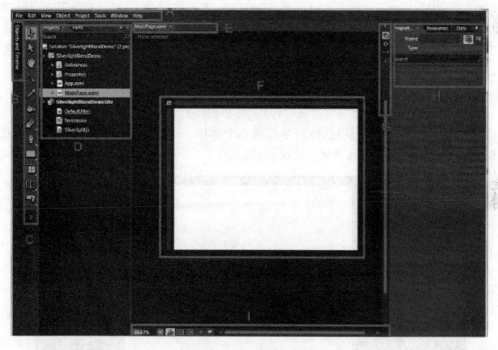

图上机 1-3 功能区介绍

创建新项目后,在 Blend 中的当前工作区,对重要的几个部分添加了标识,下面详细描述各个部分的作用。

A 部分:菜单选项。

B 部分:DockPanel 菜单。单击鼠标后,会弹出对应的窗口,例如"对象和时间线"或者"项目管理"等。

C 部分:工具面板菜单,如图上机 1-4 所示。

D 部分:项目面板,如图上机 1-5 所示。

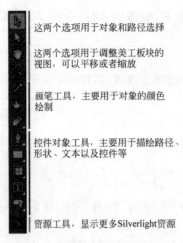

图上机 1-4　菜单按钮区

图上机 1-5　解决方案

从图上机 1-5 可以看出，在 SilverlightBlendDemo 解决方案下，有两个项目：

- SilverlightBlendDemo：该项目是 Silverlight 客户端项目，主要承载 Silverlight 客户端页面和控件。
- SilverlightBlendDemoSite：该项目是 Silverlight 服务器端项目，主要承载服务器端代码，例如 WCF Service 或 DAL 数据层代码。

E 部分：文档切换栏，该栏目显示所有打开的项目文件，可以自由切换。

F 部分：主要工作区，称为美工板，所有页面和控件设计都在该区域。

G 部分：视图和代码切换栏，该栏目提供 3 个选项，第 1 个是视图选项，第 2 个是代码选项，第 3 个是视图和代码同时显示选项，如图上机 1-6 所示。

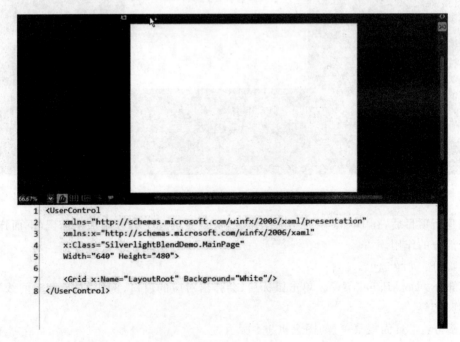

图上机 1-6　显示区

H 部分：属性和资源选项栏，从这里可以设置控件属性和对应项目资源。

I 部分：使用该部分缩放 F 部分美工板，启动动画效果，设置控件对齐选项以及查看文件注释内容。

以上窗口部件是创建项目后，默认显示的几个窗口，另外还有几个常见的窗口部件，也介绍一下：

（1）首先介绍一下扩展菜单部分，当用鼠标选中 C 部分菜单按钮，会弹出扩展菜单按钮，功能如图上机 1-7 所示。

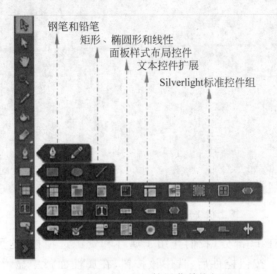

图上机 1-7　扩展菜单按钮区

（2）在扩展菜单中，最后一项是 Silverlight 资源工具按钮，单击选中后，会弹出窗口，其中包含所有控件、样式集合、行为代码集合以及动画效果和媒体文件集合。在设计时，如果添加新控件，可以从这个选项进行选择，如图上机 1-8 所示。

图上机 1-8　所有控件（资源工具）

（3）Object and Timeline（对象和时间线面板）。可以使用该面板对页面控件对象进行分层管理，另外也可以对当前对象进行动画设计，详细动画设计，将在下文描述，如图上机 1-9 所示。

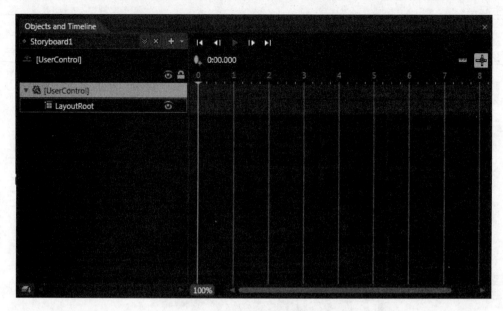

图上机 1-9　对象和时间线面板

（4）属性设置面板。在 H 部分，我们介绍了属性和资源选项栏，这里把属性栏单独列出来，在美工面板创建一个按钮控件后，选中该按钮，在属性面板中显示各种属性，例如背景、笔刷等，从这个面板可以不用输入代码，直接设计控件属性，如图上机 1-10 所示。

（5）控件模板样式资源面板。从该面板选择设计控件样式，如图上机 1-11 所示。

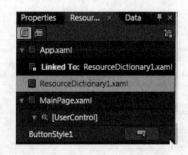

图上机 1-10　属性设置面板　　　图上机 1-11　控件模板样式资源面板（资源窗口）

(6) 调试结果面板。在该面板显示调试错误信息和输出信息，如图上机 1-12 所示。

图上机 1-12　调试结果面板（错误提示窗口）

第 2 阶段　练习

练习　使用 Blend 工具制作简单动画

问题

结合以上所学知识点，参照课本案例和指导 2 中的案例实现一个动画，要求如下：
➢ 单击按钮的时候，把矩形框的宽度变大，同时改变矩形框的位置（从上到下）。
➢ 循环次数（无限循环），效果如图上机 1-13 所示。

图上机 1-13　循环结果

上机2

矢量绘图、画刷与着色

上机任务

任务1 使用 Blend 或 VS2008 制作星光特效
任务2 使用 VS2008 制作 Silverlight 取色器
任务3 使用绘图元素绘制销售统计图形

第1阶段 指导

指导1 使用 Blend 或 VS2008 制作星光特效

完成本任务所用到的主要知识点：
- Blend 的使用
- 放射渐变画刷的使用
- 纯色画刷的使用

问题

通过前面对画刷的学习相信大家对画刷的应用有了一定的了解，那么下面通过一个案例来加强我们对画刷的理解，如图上机 2-1 所示。

分析

首先分析图上机 2-1 的组成部分。
- 两条光线：由 2 个矩形通过放射渐变画刷渐变得到。
- 光晕：由一个圆通过放射渐变画刷得到。

图上机 2-1 星光特效

解决方案

(1) 为星光设计背景特效。

```xaml
XAML:
<Canvas>
        <!-- 产生画布星光背景色 -->
        <Canvas.Background>
            <LinearGradientBrush>
                <GradientStop Color = "Blue" Offset = "0.0" />
                <GradientStop Color = "DeepSkyBlue" Offset = "1.0" />
            </LinearGradientBrush>
        </Canvas.Background>
    </Canvas>
```

(2) 下面再为这个背景添加一个星光的效果,这个效果和前面的背景一样都是使用渐变画刷实现的,这个星光效果由一个水平光线、一个垂直光线和一个放射光芒重合组成,其中这些发光效果实际上就是椭圆形加上放射性渐变产生的,完整的代码如下:

```xaml
XAML:
<Canvas>
    <!-- 产生画布星光背景色 -->
    <Canvas.Background>
        <LinearGradientBrush>
            <GradientStop Color = "Blue" Offset = "0.0" />
            <GradientStop Color = "DeepSkyBlue" Offset = "1.0" />
        </LinearGradientBrush>
    </Canvas.Background>
    <!-- 产生水平光线 -->
    <Ellipse Width = "250" Height = "10" Canvas.Left = "60" Canvas.Top = "125">
        <Ellipse.Fill>
            <RedialGradientBrush>
                <GradientStop Color = "#FFFFFFFF" Offset = "0" />
                <GradientStio Color = "#00FFFFFF" Offset = "1.0"/>
            </RedialGradientBrush>
        </Ellipse.Fill>
    </Ellipse>
    <!-- 产生垂直光线 -->
    <Ellipse Width = "10" Height = "250" Canvas.Left = "180" Canvas.Top = "20">
        <Ellipse.Fill>
            <RadialGradientBrush>
                <GradientStop Color = "#FFFFFFFF" Offset = "0" />
                <GradientStop Color = "#00FFFFFF" Offset = "1.0"/>
            <RadialGradientBrush>
        <Ellipse.Fill>
```

```
            </Ellipse>
            <!-- 产生放射光芒 -->
            <Ellipse Width="120" Height="120" Canvas.Left="125"
            Canvas.Top="70">
                <RedialGradientBrush>
                        <GradientStop Color="#FFFFFFFF" Offset="0" />
                        <GradientStop Color="#00FFFFFF" Offset="1.0"/>
                </RadialGradientBrush>
                <Ellipse.Fill>
            </Ellipse>
</Canvas>
```

指导 2 使用 VS2008 制作 Silverlight 取色器

完成本任务所用到的主要知识点：
➢ Blend
➢ Silverlight 布局
➢ 使用 C♯ 来控制画刷

问题

在上一个指导中，大家对在 XAML 语音中使用画刷有了一定的加强，那么在接下来的案例中学习如何通过 C♯ 语言来使用画刷，如图上机 2-2 所示。

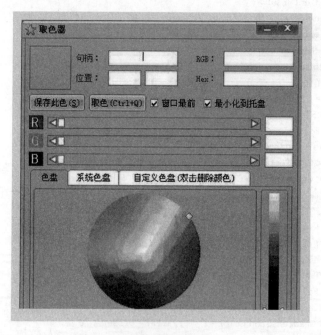

图上机 2-2 Silverlight 取色器

📝 分析

在 C# 中使用画刷时，Silverlight 默认提供的 Colors 集合只提供一部分颜色值，如果需要更多的颜色时，可以使用 Color 对象的 FromArgb 方法，FromArgb 根据传入的 RGB 和透明度的值来返回一个自定义的颜色。

使用 Grid 进行布局，用 Border 对象来产生圆角，并使用了滑动条控件来控制和产生颜色的变化，这种变化是通过在 C# 代码中声明 ValueChanged 事件来实时生成矩形的颜色值。由于 ValueChanged 事件引发的结果都是反映矩形的填充色，所以这里将滑动条的控件的 ValueChanged 事件绑定同一个事件处理程序即可。下面使用这种方法来制作一个基于 Silverlight 的取色器。

✓ 解决方案

文件代码如下：

```xml
XAML:
<Grid x:Name = "LayoutRoot" Background = "White">
    <!--添加圆角边框-->
    <Border BorderBrush = "Black" Margin = "5"
            BorderThickness = "3" CornerRadius = "10" Background = "AliceBlue">
        <!--添加网络边距-->
        <Grid Margin = "5" ShowGridLines = "False">
            <Grid.RowDefintions>
                <RowDefinition/><RowDefinition/>
                <RowDefinition/><RowDefinition/>
            <Grid.RowDefintions>
            <Grid.ColumnDefinitions>
                <ColumnDefinition/><ColumnDefinition/>
            <Grid.ColumnDefinitions>
            <!--添加滑动控制条-->
            <Slider x:Name = "sldR" Maximun = "255" Value = "255"
                Grid.Row = "0" Grid.Column = "1" />
            <Slider x:Name = "sldG" Maximun = "255" Value = "255"
                Grid.Row = "1" Grid.Column = "1" />
            <Slider x:Name = "sldB" Maximun = "255" Value = "255"
                Grid.Row = "2" Grid.Column = "1" />
            <Slider x:Name = "sldA" Maximun = "255" Value = "255"
                Grid.Row = "3" Grid.Column = "1" />
            <!--添加文字说明-->
            <TextBlock Text = "R" Width = "50" Height = "20"
                HorizontalAlignment = "Left"
                VerticalAlignment = "Top"
                Grid.Row = "0" Grid.Column = "1" />
            <TextBlock Text = "G" Width = "50" Height = "20"
                HorizontalAlignment = "Left"
                VerticalAlignment = "Top"
```

```
                    Grid.Row = "1" Grid.Column = "1" />
            <TextBlock Text = "B" Width = "50" Height = "20"
                    HorizontalAlignment = "Left"
                    VerticalAlignment = "Top"
                    Grid.Row = "2" Grid.Column = "1" />
            <TextBlock Text = "透明度" Width = "50" Height = "20"
                    HorizontalAlignment = "Left"
                    VerticalAlignment = "Top"
                    Grid.Row = "3" Grid.Column = "1" />
            <!-- 用来显示颜色 -->
            <Rectangle x:Name = "rect"
                    Width = "150" Height = "120"
                    Fill = "Green" Stroke = "Black"
                    Grid.Row = "1"
                    Grid.RowSpan = "2" Grid.Column = "0" />
            <!-- 用来输入颜色值 -->
            <TextBox x:Name = "tbColor"
                    Width = "140" Height = "26"
                    Grid.Row = "3" Grid.Column = "0" />
            <!-- 标题和说明文字 -->
            <TextBlock Text = "色值"
                    Grid.Row = "3" Grid.Column = "0" />
            <TextBlock Text = "Silverlight 取色器"
                    FontSize = "20" Width = "200" Height = "30"
                    HorizontalAlignment = "Left"
                    VerticalAlignment = "Top"
                    Grid.Row = "0" Grid.Column = "0" />
        </Grid>
    </Border>
</Grid>

C#
Public brushmcA()
{
        InitializeComponent();
        SetColors();
        //声明滑动条对象的事件处理程序
        sldR.ValueChanged +=
            new RoutedPropertyChangedEventHandler
            <double>(ValueChanged);
        sldB.ValueChanged +=
            new RoutedPropertyChangedEventHandler
            <double>(ValueChanged);
        sldG.ValueChanged +=
            new RoutedPropertyChangedEventHandler
            <double>(ValueChanged);
        sldA.ValueChanged +=
            new RoutedPropertyChangedEventHandler
            <double>(ValueChanged);
```

```
}
//填充颜色
Private void SetColors()
{
            SolidColorBrush scb = new SolidColorBrush(
                    Color.FromArgb((byte)sldA.Value,
                     (byte)sldR.Value,
                     (byte)sldG.Value,
                     (byte)sldB.Value));
            Rect.Fill = scb;
            tbColor.Text = scb.Color.ToString();
}
//滑动条的事件处理代码
Void ValueChanged(object sender,RoutedPropertyChangedEventArgs
<double> e)
{
            SetColors();
}
```

第2阶段　练习

练习　使用绘图元素绘制销售统计图形

问题

结合以上所学知识点,动手做一个统计图形。效果如图上机2-3所示。功能如下：
- 在图形底部显示月份。
- 每月的数值呈梯度上升。
- 鼠标移到每月对应的圆点的时候,提示当月的数值。

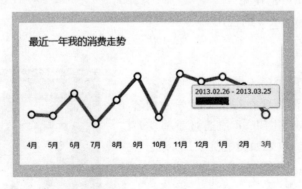

图上机2-3　显示统计图形

上机3

图像与视觉特效

上机任务

任务1　实现水中倒影效果
任务2　运用 Silverlight 3D 特性制作三维空间
任务3　Silverlight 打造特效工具栏

第1阶段　指导

指导1　实现水中倒影效果

完成本任务所用到的主要知识点：
➢ 线性渐变画刷
➢ ScaleTransform 的使用
➢ SkewTransform 的使用
➢ TranslateTransform 的使用

问题

通过前面的学习，相信大家对图像特效有了一定的知识和了解，那么现在用前面所学的内容来完成下面的内容，使用 Blend 或通过手写实现图像的水中倒影效果，效果如图上机 3-1 所示。

分析

从图上机 3-1 可以看出，以上是通过 2 张相同的图片，改变其中一张图片的缩放、扭曲和平移等这 3 种特效组合的效果。

图上机 3-1　水中倒影

解决方案

文件代码如下：

```
XAML:
<Canvas x:Name = "LayoutRoot" Background = "#000000">
    <Image Source = "../flower.jpg" Stretch = "Fill" Canvas.Top
        = "20" Canvas.Left = "100" Width = "300" Height = "200"/>
    <Image Source = "../flower.jpg" Stretch = "Fill" Canvas.Top
        = "20" Canvas.Left = "100" Width = "300" Height = "200"
        Opacity = "0.4" RenderTransformOrigin = "0,5,0.5"/>
        <Image.OpacityMask>
            <LineraGradientBrush EndPoint = "0.5,1" StartPoint
            = "0.5,0">
                <GradientStop Offset = "0.0" Color = "#00000000"/>
                <GradientStop Offset = "0.0" Color = "#FF000000"/>
            <LineraGradientBrush>
        <Image.OpacityMask>
        <Image.RenderTransform>
        <TransformGroup>
            <ScaleTransform ScaleY = "-0.75"/>
            <ScaleTransform AngleX = "-15"/>
            <Rotate Transform/>
            <TranslateTransform Y = "220" X = "-20" />
        </TransformGroup>
        </Image.RenderTransform>
    </Image>
</Canvas>
```

指导 2 运用 Silverlight 3D 特性制作三维空间

完成本任务所用到的主要知识点：
➢ Projection 使用

问题

在浏览页面时经常会发现一些 3D 特效，通过前面学习的 3D 知识来制作一个展示产品的三维空间的表单数据，如图上机 3-2 所示。

分析

本实例从表面分析可以得到通过 Projection 把 3 张图片立体化，再通过后台代码控制图片的路径达到换图片的效果。

图上机 3-2 三维空间

解决方案

(1) 文件代码如下：

```xaml
XAML:
<Grid x:Name="LayoutRoot">
    <Grid Background>
        <ImageBrush ImageSource="../Images/win7-8.jpg"/>
    <Grid Background>
    <!--图片 0-->
    <Border x:Name="r0" BorderThickness="2"
        BorderBrush="Black"
        Width="370" Height="260"
        RenderTransformOrigin="0.5,0.5">
        <Border.Background>
            <ImageBrush x:Name="img0" Stretch="Fill"/>
        </Border.Background>
        <Border.RenderTransform>
            <TransformGroup>
                <ScaleTransform x:Name="st1"/>
                <TranslateTransform x:Name="tt1" X="300"/>
            </TransformGroup>
        </Border.RenderTransform>
        </Border.Projection>
            <PlaneProjection RotationY="45"/>
        <Border.Projection>
    </Border>
    <!--图片-->
    <Border x:Name="r1" BorderThickness="2"
        BorderBrush="Black"
        Width="370" Height="260"
        RenderTransformOrigin="0.5,0.5">
        <Border.Background>
            <ImageBrush x:Name="img1" Stretch="Fill"/>
        </Border.Background>
        <Border.RenderTransform>
            <TransformGroup>
                <ScaleTransform x:Name="st0" ScaleY="0.9"/>
                <TranslateTransform x:Name="tt0"/>
            </TransformGroup>
        </Border.RenderTransform>
        <Border.Projection>
            <PlaneProjection x:Name="rt0" RotationY="0"/>
        </Border.Projection>
    </Border>
    <!--图片 2-->
    <Border x:Name="r2" BorderThickness="2">
        BorderBrush="Black"
```

```xml
            Width = "370" Height = "260"
                RenderTransformOrigin = "0.5,0.5">
    <Border.Background>
            <ImageBrush x:Name = "img2" Stretch = "Fill"/>
    </Border.Background>
    <Border.RenderTransform>
            <TransformGroup>
                    <ScaleTransform x:Name = "st2"/>
                    <TranslateTransform x:Name = "tt2" X = " – 300" />
            </TransformGroup>
    </Border.RenderTransform>
    </Border.Projection>
            <PlaneProjection x:Name = "rt2" RotationY = " – 45">
    <Border.Projection>
</Border>
<!-- 图片3 -->
<Border x:Name = "r3" BorderThickness = "2">
        BorderBrush = "Black"
        Width = "370" Height = "260"
        RenderTransformOrigin = "0.5,0.5">
    <Border.Background>
            <ImageBrush x:Name = "img3" Stretch = "Fill"/>
    </Border.Background>
    <Border.Projection>
            <PlaneProjection RotationY = " – 45"/>
    <Border.Projection>
    <Border.RenderTransform>
            <TransformGroup>
                    <ScaleTransform x:Name = "st3" ScaleY = "1.3"/>
                    <TranslateTransform x:Name = "tt3" X = " – 560" />
            </TransformGroup>
    </Border.RenderTransform>
</Border>
<!-- 播放按钮 -->
Button x:Name = "btnPlay"
Width = "70" Height = "70"
Click = "btnPlay_Click">
<!-- 定义空间模板 -->
<Button.Style>
    <Style TargetType = "Button">
            <Setter Property = "Template">
                    <Setter.Value>
                            <ControlTemplate TargetType = "Button">
                            <Image Source = "/3DPic/playoverlay.png"
                                    Width = "65" Height = "65">
                            </ControlTemplate>
                    </Setter.Value>
            </Setter>
    </Style>
```

```xml
            </Button.Style>
        </Button>
        <!-- 创建一个阴影文字显示标题 -->
        <TextBlock Text = "3DSpace Demo(Projection)">
                   HorizontalAlignment = "Left"
                   VerticalAlignment = "Top'
                   Width = "380" Height = "30"
                   Foreground = "White"
                   FontSize = "24" Margin = "5" >
            <TextBlock.Effect>
                <DropShadowEffect Color = "Black BlurRadius = "3"/>
            </TextBlock.Effect>
        </TextBlock>
</Grid>
```

以下 XAML 定义了二维空间的基本要素，那就是 4 个边框元素，每个边框都包含一个 ImageBrush 对象，那么这里为什么定义 4 个边框而不是 3 个呢？虽然界面上用户只能看到 3 幅图像，但在实际图像动画运动时，有 4 幅图像同时在场，本实例的界面完全使用了 XAML 代码来定义。

（2）XAML 用户控件资源的定义：

```xml
<UserControl.Clip>
    <RectangleGeometry Rect = "0,0 890,500" />
</UserControl.Clip>
</UserControl.Resources>
    <!--
        定义 3D 运动动画故事板
        St * :ScaleTransform 的 ScaleY 属性
        rt * :PlaneProjection 的 Rotation 属性
        tt * :TranslateTransform 的 X 属性
    -->
    <Storyboard x:Name = "myStoryboard" BeginTime = "00:00:00"
     Duration = "00:00:02">
       <DoubleAnimation Storyboard.TargetName = "st0">
                        Storyboard.TargetProperty = "ScaleY"
                        From = "0.9" To = "1" />
       <DoubleAnimation Storyboard.TargetName = "st0">
                        Storyboard.TargetProperty = "RotationY"
                        From = "0" To = "45" />
       <DoubleAnimation Storyboard.TargetName = "tt0">
                        Storyboard.TargetProperty = "X"
                        From = "0" To = "300" />
       <DoubleAnimation Storyboard.TargetName = "st1">
                        Storyboard.TargetProperty = " ScaleY"
                        From = "1" To = "1.3" />
       <DoubleAnimation Storyboard.TargetName = "tt0">
                        Storyboard.TargetProperty = "X"
```

```
                    From = "300" To = "560" />
        <DoubleAnimation Storyboard.TargetName = "st2">
                    Storyboard.TargetProperty = " ScaleY"
                    From = "1" To = "0.9" />
        <DoubleAnimation Storyboard.TargetName = "tt2">
                    Storyboard.TargetProperty = "RotationY"
                    From = " - 45" To = "0" />
        <DoubleAnimation Storyboard.TargetName = "st3">
                    Storyboard.TargetProperty = "ScaleY"
                    From = "1.3" To = "1" />
        <DoubleAnimation Storyboard.TargetName = "tt3">
                    Storyboard.TargetProperty = "X"
                    From = " - 560" To = " - 300" />
</Storyboard>
</UserControl.Resources>
```

有了边框和边框中的图像,那么剩下的就是动画的运动效果了,以上 XAML 完成了这一功能,动画的作用属性和作用目标是针前面界面中的边框元素的变形属性。

本实例的界面边框和图片等元素完全可以使用 C#代码来动态生成,这样看起来更简洁和灵活,有兴趣的读者可以进一步优化这个实例。

(3) 使用 C#代码来动态生成界面边框和图片等元素:

```
Public OpenPage3D()
{
        InitializeComponent();
        This.Loaded += new RoutedEventHandler(OpenPage3D_Loaded);
}
Private void OpenPage3D_Loaded(object sender,RoutedEventArgs e)
{
    //初始化位置
    CurPos = 0;
    SetEasingFunction();
    SetSource();
}
Private void btnPlay_Click(Object sender,RoutedEventArgs e)
{
    PlayStory();
}
//播放事件的方法
Private void PlayStory()
{
    SetSource();
    //播放动画
    myStoryboard.Begin();
    AddNum();
}
//改变位置
```

```
Private void AddNum()
{
    If(CurPos < 10)
    {
       CurPos++;
    }
    Else
    {
     CurPos = 0;
    }
}
//当前位置
Private int CurPos
{
    Get;
    Set;
}
//获取图片源文件
Private ImageSource GetSource(int num)
{
    ImageSource source = new BitmapImage(
       new Uri("/3DPic/" + num + ".jpg",UriKind.Relative));
    return source;
}
//设置当前位置的图片源
Private void SetSource()
{
    Img0.ImageSource = GetSource(CurPos);
    Img1.ImageSource = GetSource(CurPos + 1);
    Img2.ImageSource = GetSource(CurPos + 2);
    Img3.ImageSource = GetSource(CurPos + 3);
}
//使用 Silverlight3 的 EasingFunction 动画
Private void SetEasingFunction()
{
    //创建 EasingFunction 动画中的 CircleEase 对象
    CircleEase ce = new CircleEase();
    Ce.EasingMode = EasingMode.EaseOut;
    //遍历故事板中的所有动画
    For( ing i = 0;i < myStoryboard.Children.Count;i++ )
    {
       DoubleAnimation da = myStoryboard.Childeren[i] as
       DoubleAnimation;
       If(null!= da)
       {
          //设置 DoubleAnimation 的 EasingFunction 属性
          Da.EasingFunction = ce;
       }
    }
}
```

实例中的 SetSource 函数根据当前位置变量来调用 GetSource 获取图片源。

第 2 阶段 练习

练习 运用 Silverlight 打造特效工具栏

问题

结合以上所学知识点,参照上面的案例实现特性工具栏。效果如图上机 3-3 所示。要求如下:

➢ 显示相关的所有选项。
➢ 鼠标移动到小图标时显示该图标的大尺寸图标。
➢ 鼠标离开图标则图标还原成小图标。

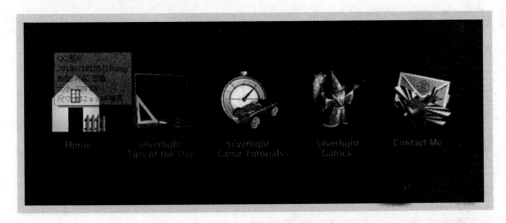

图上机 3-3 特效工具栏

首先添加一个 UserAction 的类,有一个处理用户登录的 doLogin 方法,配置 Struts 2 框架时需要两个配置文件,一个是 web.xml,另一个就是 struts.xml。

上机 4

动画与多媒体

上机任务

任务 1　结合动画与控件开发跑马灯图片浏览器
任务 2　全功能视频播放器
任务 3　制作 Silverlight 时钟效果

第 1 阶段　指导

指导 1　结合动画与控件开发跑马灯图片浏览器

涉及重要知识点：
- Blend 的基本用法
- C#管理动画
- DoubleAnimation 线性动画

问题

前面学习了动画的一些基本用法，现在把前面的知识结合起来做一个跑马灯图片浏览器，效果如图上机 4-1 所示。

图上机 4-1　跑马灯效果

分析

跑马灯动画是在网页中常见的动画效果，它可以在固定大小的范围显示更多的内容，一般来说跑马灯效果主要是应用在文字内容上，使其产生自上而下或自左而右的循环运动，不过本实例的 Silverlight 跑马灯动画则是采用图片来代替文字，通过 Silverlight 的动画属性驱动一组产品介绍图片元素，使它们产生水平运动，并且这种运动是循环不间断的。

在这个实例中主要实现了以下功能：

- 使用动画属性驱动图片的运动动画。
- 图片远动后最后一幅会自动循环。
- 单击查看放大图片。
- 当鼠标放到图片时运动的图片会停止，当鼠标离开时暂停的图片会继续运动。
- 当鼠标单击任何一个图片时，该图片会显示真正大小。

解决方案

（1）主界面：

```xml
<Grid x:Name = "Layout" Background = "White">
    <Canvas x:Name = "canvas" Background = "Black" Grid.Row = "1"
        Height = "280">
        <!-- 隐藏矩形以外的其他部分 -->
        <Canvas.Clip>
            <RectangleGeometry x:Name = "rg" />
        </Canvas.Clip>
        <StackPanel x:Name = "sp" Orientation = "Horizontal" >
        </StackPanel>
    </Canvas>
    <Image x:Name = "img_Full" Width = "640" Height = "480"
        Visibility = "Collapsed"
        MouseLeftButtonUp = "img_Full_MouseLeftButtonUp" />
</Grid>
```

界面由 Grid、Canvas、StackPanel 和一个 Image 组成，Image 用来显示图片的真实尺寸。下面是 C# 代码：

```csharp
C#
public partial class Demo : UserControl
{
    //定义常量和变量
    private Storyboard storyboard;
    private const double photowidth = 320;
    private double totalwidth;
    public Demo()
    {
        InitializeComponent();
        CreatePhoto();
```

```csharp
        }
        //创建图片列表
        private void CreatePhoto()
        {
            string[] picList = new string[] { "1.jpg", "2.jpg",
                "3.jpg","4.jpg", "5.jpg" };
            //创建多组图片,保证图片不会出现空白,因为 StackPanel 是横向排列的,
            //将图片排成一圈
            for (int i = 0; i < 3; i++)
            {
                //根据数组创建图片
                for (int j = 0; j < picList.Length; j++)
                {
                    UC_pic pic = new UC_pic();
                    pic.ImageUrl = "../images/photo/" + picList[j];
                    pic.Width = photowidth;
                    //绑定事件
                    pic.MouseEnter +=
                        new MouseEventHandler(pic_MouseEnter);
                    pic.MouseLeave +=
                        new MouseEventHandler(pic_MouseLeave);
                    pic.MouseLeftButtonUp +=
                        new MouseButtonEventHandler(pic_MouseLeftButtonUp);
                    //添加对象到 StackPanel 中
                    sp.Children.Add(pic);
                }
            }
            //计算图片的总宽度
            totalwidth = -1.0 * photowidth * picList.Length;
            Canvas.SetLeft(sp, totalwidth);
            //调用初始化方法
            CreateStoryboard();
            //播放动画
            storyboard.Begin();
            //重新绘制区域
            Resize();
        }
        //创建故事面板
        private void CreateStoryboard()
        {
        //创建故事面板
        storyboard = new Storyboard();
        DoubleAnimation animation = new DoubleAnimation();
        //设置动画延时
        animation.Duration =
         new Duration(TimeSpan.FromSeconds(2.0));
        //设置对象的作用属性
        Storyboard.SetTarget(animation, sp);
```

```csharp
    Storyboard.SetTargetProperty(animation,
     new PropertyPath("(Canvas.Left)", new object[0]));
    //添加到动画故事板内
    storyboard.Children.Add(animation);
    //动画自动完成事件
    storyboard.Completed +=
     new EventHandler(storyboard_Completed);
}
//动画自动完成事件,当动画播放完成(结束)时,再次循环动画
void storyboard_Completed(object sender, EventArgs e)
{
    DoubleAnimation animation =
    (DoubleAnimation)storyboard.Children[0];
    //取得图片当前位置
    double left = Canvas.GetLeft(sp);
    //如果图片已接近最后,就重新设置位置
    if (left > (totalwidth - photowidth))
    {
        animation.From = new double?(left);
    }
    //设置动画的起始值(From)所依据的总量(总长度)
    animation.By = new double?(totalwidth);
    //循环动画
    storyboard.Begin();
}
private void Resize()
{
    //重新绘制显示区域
    rg.Rect = new Rect(0, 0, this.ActualWidth, 260);
}
void pic_MouseLeftButtonUp(object sender,
    MouseButtonEventArgs e)
{
    //显示放大图片
    UC_pic pic = sender as UC_pic;
    img_Full.Source = pic.photo.Source;
    img_Full.Visibility = Visibility.Visible;
}
void pic_MouseLeave(object sender, MouseEventArgs e)
{
    //继续动画
    storyboard.Resume();
}
void pic_MouseEnter(object sender, MouseEventArgs e)
{
    //暂停动画
    storyboard.Pause();
}
private void img_Full_MouseLeftButtonUp(object sender,
```

```
            MouseButtonEventArgs e)
        {
            //隐藏放大图片
            img_Full.Visibility = Visibility.Collapsed;
        }
        private void UserControl_SizeChanged(object sender,
            SizeChangedEventArgs e)
        {
            //动画根据屏幕大小改变而改变
            Resize();
        }
    }
```

C#代码部分的主要动作是动态创建故事板并把故事板动画应用到相对应的 UC 上。需要注意的是,这里运动的图片并不是一个 Image 对象,而是一个名为 UC 的 XAMLUserControl。相比简单的 Image 元素来说,使用 UC 作为运动单位有更好的可扩展性。可以在这个 UserControl 中添加更丰富的内容,不管需要显示多少图片,只需要一个 XAML 文件即可,UC 的代码如下:

```
<Canvas x:Name = "LayoutRoot" Background = "White">
    <Image x:Name = "photo" Width = "320" Height = "240"
        Stretch = "UniformToFill" Margin = "10" />
</Canvas>
```

(2) UC.Xaml:

```
public partial class UC_pic : UserControl
    {
        public UC_pic()
        {
            InitializeComponent();
        }
        //定义一个 imageUrl 属性
        private string _imgUrl;
        public string ImageUrl
        {
            get { return this._imgUrl; }
            set {
                //设置图片资源属性
                this._imgUrl = value;
                Uri uri = new Uri(value, UriKind.Relative);
                BitmapImage bitimg = new BitmapImage(uri);
                this.photo.Source = bitimg;
            }
        }
    }
```

指导2 全功能视频播放器

完成本任务所用到的主要知识点：
➢ MediaElement 对象的使用
➢ 全屏支持模式

问题

前面介绍视频的内容仅是实现视频和音频的最简单的直播，实际应用中往往需要添加对多媒体内容的实时控制功能。下面制作一个全功能播放器，效果如图上机 4-2 所示。

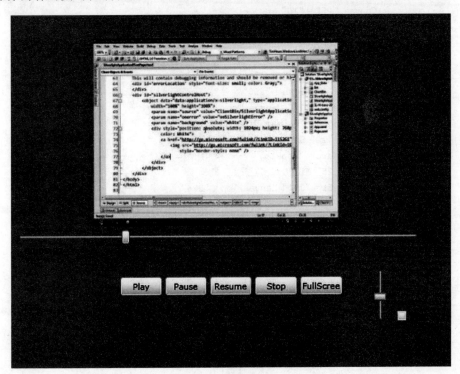

图上机 4-2 全功能播放器

分析

MediaElement 为此提供了十分方便的操作方法，上面案例不仅实现了对视频内容的基本控制功能，还支持视频播放状态与时间的显示，对视频播放进度的显示和控制，以及全屏播放功能。

解决方案

（1）设计主页面：

```
<UserControl
    xmlns = "http://schemas.microsoft.com/winfx/2006/xaml/presentation"
```

```xml
xmlns:x = "http://schemas.microsoft.com/winfx/2006/xaml"
x:Class = "VideoPaly.MainPage"
Width = "640" Height = "480">
< Grid x:Name = "LayoutRoot" Background = "Gray">
    < Canvas Margin = "34,23,38,63">
        < MediaElement x:Name = "media"
            Source = "http://download.microsoft.com/download/2/0/5/
            205d8c39-3d55-4032-8195-7b0e6eda4cb6/WinVideo-SL-
            InstallExperience.wmv"
            Height = "248"
            Width = "542"
            Canvas.Left = "10"
            Canvas.Top = "8"
            BufferingProgressChanged = "media_BufferingProgressChanged"
            Loaded = "media_Loaded"/>
        < Slider x:Name = "slider1"
            Height = "21" Width = "550"
            Canvas.Left = "10" Canvas.Top = "260"
            ValueChanged = "Slider_ValueChanged"
            Maximum = "10" SmallChange = "1"/>
        < Button x:Name = "btnPlay"
            Height = "25" Width = "57"
            Canvas.Left = "151" Canvas.Top = "324"
            Content = "Play"
            Click = "btnPlay_Click"/>
        < Button x:Name = "btnPause"
            Height = "25" Width = "57"
            Canvas.Left = "212" Canvas.Top = "324"
            Content = "Pause"
            Click = "btnPause_Click"/>
        < Button x:Name = "btnResume"
            Height = "25" Width = "57"
            Canvas.Left = "273" Canvas.Top = "324"
            Content = "Resume"
            Click = "btnResume_Click"/>
        < Button x:Name = "btnStop"
            Height = "25" Width = "57"
            Canvas.Left = "334" Canvas.Top = "324"
            Content = "Stop"
            Click = "btnStop_Click"/>
        < TextBlock x:Name = "tbCurrentTime"
            Height = "25" Width = "90"
            Canvas.Left = "22" Canvas.Top = "281"
            TextWrapping = "Wrap"/>
        < TextBlock x:Name = "tbTotalTime"
            Height = "25" Width = "91"
            Canvas.Left = "505" Canvas.Top = "281"
            TextWrapping = "Wrap"/>
        < CheckBox x:Name = "cbSilent"
```

```
                    Canvas.Left = "526" Canvas.Top = "369"
                    Content = "静音"
                    RenderTransformOrigin = "0.881,0.235"
                    Checked = "cbSilent_Checked"
                    Unchecked = "cbSilent_Unchecked"/>
                <TextBlock x:Name = "tbBufferingdValue"
                    Height = "55" Width = "171"
                    Canvas.Left = "213"
                    Canvas.Top = "111"
                    TextWrapping = "Wrap"
                    FontSize = "18.667"/>
                <Slider x:Name = "sliderVolumn"
                    Height = "74" Width = "36"
                    Canvas.Left = "486"
                    Canvas.Top = "312"
                    Orientation = "Vertical"
                    ValueChanged = "sliderVolumn_ValueChanged"
                    Maximum = "1" LargeChange = "0.2"/>
                <Button x:Name = "btnFullScreen"
                    Height = "25" Width = "57"
                    Canvas.Left = "395" Canvas.Top = "324"
                    Content = "FullScreen"
                    Click = "btnFullScreen_Click"/>
        </Canvas>
    </Grid>
</UserControl>
```

（2）使用 CurrentStateChanged 事件，可以使用它来得到 MediaElement 对象的当前状态，然后是对播放器按钮事件的处理，代码如下：

```
using System;
using System.Windows;
using System.Windows.Controls;
using System.Windows.Documents;
using System.Windows.Ink;
using System.Windows.Input;
using System.Windows.Media;
using System.Windows.Media.Animation;
using System.Windows.Shapes;
using System.Windows.Threading;
namespace VideoPaly
{
    public partial class MainPage : UserControl
    {
        //使用计时器对象来更新播放进度
        private DispatcherTimer timer = new DispatcherTimer();
        //媒体的时长
        private TimeSpan duration;
```

```csharp
public MainPage()
{
    //为初始化变量所必需
    InitializeComponent();
}
private void btnPlay_Click(object sender,
  System.Windows.RoutedEventArgs e)
{
    //播放
    media.Play();
}
private void btnPause_Click(object sender,
  System.Windows.RoutedEventArgs e)
{
    //暂停
    media.Pause();
}
private void btnResume_Click(object sender,
  System.Windows.RoutedEventArgs e)
{
    //继续
    media.Play();
}
private void btnStop_Click(object sender,
   System.Windows.RoutedEventArgs e)
{
    //停止
    media.Stop();
}
private void cbSilent_Checked(object sender,
  System.Windows.RoutedEventArgs e)
{
    //静音
    media.IsMuted = true;
}
private void cbSilent_Unchecked(object sender,
 System.Windows.RoutedEventArgs e)
{
    //不静音
    media.IsMuted = false;
}
private void Slider_ValueChanged(object sender,
 System.Windows.RoutedPropertyChangedEventArgs<double> e)
{
    //确定多媒体对象的播放位置是否处于可管理状态
    if (media.CanSeek)
    {
        //重新设置 MediaElement 的播放位置
```

```csharp
                    media.Position =
                        TimeSpan.FromSeconds(
                            media.NaturalDuration.TimeSpan
                            .TotalSeconds * slider1.Value)/10);
                }
            }
            private void media_BufferingProgressChanged(object sender,
System.Windows.RoutedEventArgs e)
            {
                //视频缓冲事件
                double bufferingValue = this.media.BufferingProgress * 100;
                this.tbBufferingValue.Text = "缓冲:"
                    + bufferingValue.ToString() + "%";
                if(this.tbBufferingValue.Text == "缓冲:100%")
                {
                    //隐藏缓冲值
                    this.tbBufferingValue.Visibility =
                        Visibility.Collapsed;
                }
            }
            private void media_Loaded(object sender,
              System.Windows.RoutedEventArgs e)
            {
                //设置事件引发的时间间隔
                timer.Interval = TimeSpan.FromMilliseconds(500);
                //计时器对象事件
                timer.Tick += new EventHandler(timer_Tick);
                //开始计时
                timer.Start();
                //设置音量
                media.Volume = 0.8;
                sliderVolumn.Value = 0.8;

            }
            private void timer_Tick(object sender, System.EventArgs e)
            {
                //当前处于播放时有效
                if (this.media.CurrentState == MediaElementState.Playing)
                {
                    duration = media.NaturalDuration.HasTimeSpan ?
                        media.NaturalDuration.TimeSpan :
                        TimeSpan.FromMilliseconds(0);
                    this.tbCurrentTime.Text = string.Format("{0}:{1}:{2}",
                        media.Position.Hours < 10 ? "0"
                        + media.Position.Hours.ToString()
                        : media.Position.Hours.ToString(),
                        media.Position.Minutes < 10 ? "0"
                        + media.Position.Minutes.ToString()
                        : media.Position.Minutes.ToString(),
```

```
                    media.Position.Seconds < 10 ? "0"
                        + media.Position.Seconds.ToString()
                        : media.Position.Seconds.ToString());
                this.tbTotalTime.Text = string.Format(
                "{0}{1:00}:{2:00}:{3:00}", "时长:",
                duration.Hours,
                duration.Minutes,
                duration.Seconds);
                //跳过事件处理程序
                slider1.ValueChanged -=
                    new RoutedPropertyChangedEventHandler<double>
                    (Slider_ValueChanged);
                //计算并设置 Slider 的百分比
                slider1.Value = (media.Position.TotalSeconds
                    //media.NaturalDuration.TimeSpan.TotalSeconds) * 10;
                //重新声明 ValueChanged 事件
                slider1.ValueChanged +=
                    new RoutedPropertyChangedEventHandler<double>
                    (Slider_ValueChanged);
            }
        }
        private void sliderVolumn_ValueChanged(object sender,
           System.Windows.RoutedPropertyChangedEventArgs<double> e)
        {
            //音量发生改变触发事件
            media.Volume = sliderVolumn.Value;
        }
        private void btnFullScreen_Click(object sender,
         System.Windows.RoutedEventArgs e)
        {
            //全屏操作
            Application.Current.Host.Content.IsFullScreen
                = !Application.Current.Host.Content.IsFullScreen;
        }
    }
}
```

第 2 阶段 练习

练习 制作 Silverlight 时钟效果

问题

结合以上所学知识以及 Blend 工具,使用自定义依赖项属性,制作一个 Silverlight 时钟。效果如图上机 4-3 所示。功能要求如下:

- 显示当前时间。
- 使用样式。
- 每一秒,时间改变并且有翻动的效果。

图上机 4-3　特效时钟

上机 5

Silverlight 与 HTML、JavaScript 三者交互

上机任务

任务 1　创建一个 Silverlight 程序
任务 2　保存 Cookie
任务 3　读取 Cookie 文件里面的用户名和密码实现自动登录

第 1 阶段　指导

指导 1　创建一个 Silverlight 程序

完成本任务所用到的主要知识点：
➢ Silverlight 布局

问题

实例 1：创建这章的学习，结合以前的知识，首先来创建一个登录的页面，效果如图上机 5-1 所示。

图上机 5-1　登录页面

分析

从上图分析页面包含 2 个 TextBlock、2 个文本框和 2 个 Button。

解决方案

(1) 创建程序。

(2) 登录页面代码如下:

```xml
<Grid x:Name = "LayoutRoot" Background = "White">
  <Grid.ColumnDefinitions>
    <ColumnDefinition Width = "100"/>
    <ColumnDefinition Width = "300"/>
  </Grid.ColumnDefinitions>
<Grid.RowDefinitions>
    <RowDefinition Height = "50"/>
    <RowDefinition Height = "50"/>
    <RowDefinition Height = "50"/>
    <RowDefinition Height = "150 *"/>
</Grid.RowDefinitions>
<TextBlock Name = "uersName"Text = "用户名:"Grid.Column = "0"Gird.Row = "0"
    HorizontalAlignment = "Center" VertivalAlignment = "Center"/>
<TextBlock Name = "userPwd" Text = "密码:"Grid.Column = "0"Grid.Row = "1"
    HorizonatalAlignment = "Center"VerticalAlignment = "Center"/>
<TextBox Name = "txtUserName" Width = "200" Height = "40"Grid.column = "1"

Grid.Row = "0"HorizontalAlignment = "Center"VerticalAlignment = "Center"/>
  <TextBox Name = "txtUserPwd"Width = "200"Height = "40"Grid.Column = "1"
Grid.Row = "1"HorizontalAlignment = "Center"VerticalAlignmnt = "Center">
  <Button Name = "ensure"Content = "确定"Width = "60"Height = "30"Grid.Row = "2"
    Margin = "0,10,240,10"Grid.Column = "1"/>
  <Button Name = "Cancel"Content = "取消"Width = "60"Height = "30"Grid.Row = "2"
    Margin = "141,10,99,10"Grid.Column = "1"/>
</Grid>
```

指导 2　保存 Cookie

完成本任务所用到的主要知识点:
- Silverlight 与 HTML 交互
- Silverlight 操作 Cookie

问题

实例 2:我们都知道在 ASP.NET 中,Cookie 的作用是保存少量的数据客户端,同样 Silverlight 也可以操作 Cookie 来保存信息到客户端,现在在实例 1 的基础上,把用户登录的用户名和密码发到客户端,如图上机 5-2 所示。

图上机 5-2　保存 Cookie

📝 分析

本例主要是通过 Silverlight 操作 Cookie 发送到客户端。

✅ 解决方案

文件代码如下：

```
CookieHelper
Public static void SetCookie(string userName, string userPwd)
{
DateTime expiration = DateTime.UtcNow + TimeSpan.FromDays(2000);
String cookie = String.Format("userName = {0};expires = {1}",username,
    expiration.ToString("R"));
String cookie = String.Format("userPwd = {0};expires = {1}",userPwd,
    expiration.ToString("R"));
HtmlPage.Document.SetProperty("Cookie",cookie);
HtmlPage.Document.SetProperty("Cookie",cookie1);
}
```

第 2 阶段　练习

练习　读取 Cookie 文件中的用户名密码实现自动登录

❓ 问题

实例 3：结合以上所学知识点，Silverlight 操作 Cookie，改进前面的案例。效果如图上机 5-3 所示。

验证要求如下：
- ➢ 判断用户名和密码不能为空。
- ➢ 判断保存用户名和密码到 Cookie 文件中，并发送到客户端。
- ➢ 在下次登录时自动获取客户端 Cookie 文件中的用户名和密码，显示登录成功提示框，不需要跳转。

上机5　Silverlight与HTML、JavaScript三者交互

用户名：
密　码：
确认　　取消

图上机 5-3　提示效果

可以参考本章理论课中的例子。

上机 6

数据访问与 Silverlight 高级应用实例

上机任务

任务 1　实现用户登录
任务 2　实现员工管理的新增
任务 3　实现员工管理的删除和修改

第 1 阶段　指导

指导 1　实现用户登录

完成本任务所用到的主要知识点：
➤ WCF 的使用
➤ Linq To Sql 的使用
➤ 页面的跳转

问题

前面上机 5 的练习中已经做过登录，只是没有涉及与数据交互，现在通过 WCF 服务来实现用户的登录，效果如图上机 6-1 所示。

分析

图 6-1 中要实现与数据进行交互，首先要创建员工表（empInfo）、角色表（roleInfo）、WCF 服务

图上机 6-1　登录页面

（LoginServer.svc）以及登录页面（Login.xaml）。

✅ 解决方案

(1) 在 SQL Server Management Studio 中创建数据库以及数据库表。

```
Use master
Go
If exists(select 1 from sys.sysdatabase where name = 'DressDB')
Drop database DressDB
Go

Create database DressDB
On primary
(
    Name = "DressDB_data",
    Filename = 'D:\学习资料\DressDB_data.mdf',
    Size = 10,
    Filegrowth = "30%"
)
Log on
(
    Name = "DressDB_log",
    Filename = 'D:\学习资料\DressDB_log.ldf',
    Size = 3,
    Filegrowth = "10%"
)
Go
Use DressDB
Go
--商品表
Create table shopInfo
(
sNo varchar(10)primary key, --商品编号
sName varchar(10)not null, --商品名称
Price money not null, --价格
)
Go
Select * from shopInfo
Go
--角色表
Create table roleInfo
(
  roNo varchar(20)primary key, --角色编号
  roName varchar(20)unique not null, --角色名称
)
Go
--权限表
Create table rightInfo
```

```sql
(
    riNo varchar(10)primary key , -- 权限编号
    riName varchar(100)not null, -- 权限描述
    parentNo varchar(10), -- 父权限编号
    Uri varchar(1000), -- 权限对应页面
)
Go
-- 角色权限关联表
Create table rrInfo
(
 rrNo int primary key identity(1,1), -- 角色权限关联项编号
 roNo varchar(20)references roleInfo(roNo), -- 外键:角色编号
 riNo varchar(10)references rightInfo(riNo) -- 外键:权限编号
)
Go
-- 员工表
Create table empInfo
(
 eNo varchar(10)primary key, -- 员工编号
 eName varchar(20)not null, -- 员工姓名
 Pwd varchar(50)not null, -- 密码
 eRoleNo varchar(20)not null references roleInfo(roNo), -- 外键(角色编号)
 State varchar(10)not null -- 员工状态(在,离职)
)
Go -- 订单表
Create table orderInfo
(
 oNo varchar(20)primary key, -- 订单编号
 oDate datetime not null, -- 订单日期
 oAmount int not null, -- 总数量
 oPrice money not null, -- 总价
 oEempNo varchar(10)not null references empInfo(eNo), -- 外键员工编号
 Remark varchar(200) -- 备注:折扣信息
)
Go
Select * from orderInfo
Go
-- 详细订表单
Create table orderDetail
(
 odNo varchar(20)primary key, -- 详细订单编号
 odSno varchar(10)not null references shopInfo(sNo), -- 外键:商品编号
 odONo varchar(20)not null references orderInfo(oNo), -- 外键:订单号
 odAmount int not null, -- 商品数量
 OdPrice money not null, -- 总价
)
Go
Select * from orderDetail
Go
```

```sql
-- 促销方式字典
Create table saleModeDictionary
(
  smdNo int primary key identity(1,1), -- 促销方式编号
  smdValue varchar(200)not null, -- 促销方式表述
  State varchar(10)not null default("禁用") -- (启用,禁用)
)
Go
-- 存储过程:根据角色查询权限
Create procedure proc_getRightByRoNo
@no varchar(20)
As
 Select ri.riNo,ri.riName,ri.parentNo,ri.uri from roleInfo ro
Inner join rrInfo rr on rr.roNo = ro.roNo
Inner join rightInfo ri on rr.riNo = ri.riNo
Where ro.roNo = @no
Go
Execute proc_getRightByRoNo 'admin'
GO
-- 存储过程:查询某个角色没有的权限
Create procedure proc_getNoRightByRoNo
@no varchar(20)
As
Select riNo,riName,parentNo,uri from rightInfo where riNo notin(
 Select ri.riNo from roleInfo ro
 Inner join rrInfo rr on rr.roNo = ro.roNo
 Inner join rightInfo ri on rr.riNo = ri.riNo
Where ro.roNo = @no
)
Go
Execute proc_getNoRightByRoNo 'admin'
Go
-- 存储过程:查询用户权限
Create procedure proc_getRight
@name varchar(20)
As
Select ri.riNo,riName,ri.parentNo,ri.uri from empInfo emp
 Inner join roleInfo ro on emp.eRoleNo = ro.roNo
 Inner join rrInfo rr on rr.roNo = ro.roNo
 Inner join rightInfo ri on ri.riNo = ri.riNo
 Where emp.eNo = @name

go
 Exec procedure proc_Tongji
 @year varchar(4)
 @month varchar(2)
 @day varchar(2)
As
 If@month = "and@day" = "and@year = "
```

```sql
Begin
 Select'店铺'as'type',ISNULL(SUM(oPrice),0)as'totalPrice'from
 orderInfo
End
 --按年查
Else if @month = "and@day = "and@year!= "
Begin
Select'店铺'as'type',ISNULL(SUM(oPrice),0)as'totalPrice'from
orderInfo
End
 --按月查
Else if@day = "and@year!= "
 Begin
 Select'店铺'as'type',ISNULL(SUM(oPrice),0)as'totalPrice'from
 orderInfo where
CONVERT(varchar,DATEPART(YEAR,oDate)) = @year and
CONVERT(varchar,DATEPART(mobth,oDate)) = @month
End
 --按天查
Else
Begin
   Select '店铺'as'type',ISNULL(SUM(oPrice),0)as'totalPrice'from
    orderInfo where
CONVERT(varchar,DATEPART(YEAR,oDate)) = @year and
CONVERT(varchar,DATEPART(month,oDate)) = @month and
CONVERT(varchar,DATEPART(DAY,oDate)) = @day
End
Go
Execute proc_Tongji",","
Go
 --个人
Create procedure proc_EempTongJi
@empNo varchar
@year varchar(4)
@month varchar(2)
@day varchar(2)
As
 If@month = "and@day = "and@year = "
 Begin
 Select @empNo as'type',ISNULL(SUM(oPrice),0)as'totalPrice'from
   orderInfo group by oEmpNo having oEmpNo = @ = empNo
 End
 --按年
Else if@month = "and@day = "
 Begin
 Select @empNo as'type',ISNULL(SUM(oPrice),0)as'totalPrice'from
 orderInfo where CONVERT(varchar,DATEPART(YEAR,oDate)) = @year
   Group by oEmpNo having oEmpNo = @empNo
End
```

```
--按月
Else if @day = ''
Begin
  Select @empNo as 'type', ISNULL(SUM(oPrice),0) as 'totalPrice' from
  orderInfo where CONVERT(varchar,DATEPART(YEAR,oDate)) = @year
  And CONVERT(varchar,DATEPART(month,oDate)) = @month
  Group by oEempNo having oEmpNo = @empNo
End
--按天
Else
Begin
  Select @empNo as 'type', ISNULL(SUM(oPrice),0) as 'totalPrice'
  from orderInfo
  Where CONVERT(varchar,DATEPART(YEAR,oDate)) = @year
  AndCONVERT(varchar,DATEPART(month,oDate)) = @month
  And CONVERT(varcar,DATEPART(DAY,oDate)) = @day
  Group by oEmpNo having oEmpNo = @empNo
End
Go
Execute proc_EempTongJi 'e01','2011','2','25'
Go
Execute proc_EmpTongJi 'e01','',''
go
```

（2）从上面的内容发现了与登录无关的表和存储过程，这些是为下面的练习做准备的。现在创建实体数据模型（DressMSModel.edmx），如图上机 6-2 所示。

图上机 6-2　创建实体数据模型（DressMSModel.edmx）

（3）创建 WCF 服务（LoginService.svc）如图上机 6-3 所示。

在 WCF 服务中，添加一个登录的方法（Login）。方法有 2 个参数：用户名、密码，代码如下：

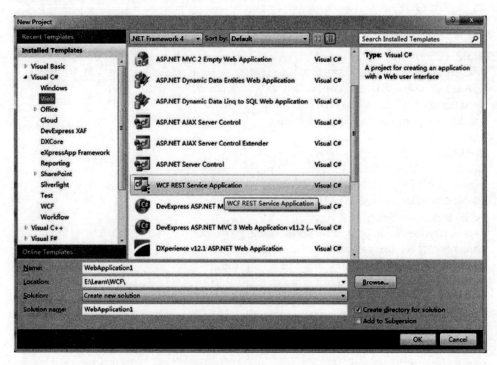

图上机 6-3　创建 WCF 服务（LoginService.svc）

```
//MD5 加密
//公共方法：获取给定字符的 MD5 码
Public string GetHashCode(string input)
{
 //返回的字符串
 StringBuilder sb = new StringBuilder();
 //创建一个 MD5 示例
 MD5 md5Hasher = MD5.Create();
 //把传入的密码编码为字节序列
 Byte[ ]data = Encoding.Default.GetBytes(input);
 //获得字节序列的编码
 Byte[ ]hasher = md5Hasher.ComputeHash(data);
 Foreach(var myByte in hasher)
 {
  sb.Append(myByte.ToString());
 }
 Return sb.ToString();
}
DressDBEntities entities = new DressDBEntities();
//在此处添加更多操作并使用[OperationContract]标记它们
//登录
 [OperationContract]
Public int Login(string name,string pwd)
{
```

```
//获取密码的哈希码
String input = GetHashCode(pwd);
Var users = from emp in entities.empInfo where emp.eNo == name
 &&emp.pwd == input&&emp.state == "在职"select emp;
Return users.Count();
}
```

(4) 设置登录页面(Login.xaml)。

```xml
<Grid x:Name = "LayoutRoot"Background = "Black">
 <!-- 背景 -->
 <!--<Grid.Background>
     <ImageBrush ImageSource = "Image/bg.jpg"/>
 </Grid.Background>-->
 <!-- 登录框 -->
 <Border x:Name = "dbLogin" CornerRadius = "10"Width = "250"Height = "150">
<Grid>
<Grid.Background>
<ImageBrush ImageSource = "Imgs/bg_login_pane.jpg"/>
</Grid.Background>
<Grid.RowDefinitions>
<RowDefinition Height = "*"/>
<RowDefinition Height = "*"/>
<RowDefinition Height = "*"/>
<RowDefinition Height = "*"/>
<RowDefinition Height = "0.5*"/>
</Grid.RowDefinitions>
<Grid.ColumnDefinitions>
<ColumnDefinition Width = "0.5*"/>
<ColumnDefinition Width = "*"/>
<ColumnDefinition Width = "*"/>
<ColumnDefinition Width = "0.5*"/>
<Grid.ColumnDefinitions>
 <TextBlock Grid.Row = "1"Grid.Column = "1"Text = "编号:"FontSize = "14"
   VerticalAlignment = "Center"HorizontalAlignment = "Center"/>
 <TextBox x:Name = "txtName" Text = "e01"Hright = "23"Grid.Row = "1"
   Width = "90"Grid.Column = "2"/>
 <TextBlock Grid.Row = "2"Grid.Column = "1" Text = "密码:"FontSize = "14"
   VerticalAlignment = "Center"HorizontalAlignment = "Center"/>
 <passwordBox x:Name = "txtPwd"Password = "123"Height = "23"Grid.Row = "2"
   Grid.Column = "2"MaxLength = "6"Width = "90"/>
 <Button Content = "登录"Grid.Row = "3"Grid.Column = "1" Click = "btnOK_Click"
   HorizontalAlignment = "Center"Width = "75"/>
 <Button Content = "重置"Grid.Row = "3"Grid.Column = "2"
   Click = "btnReset_Click" Height = "23" HorizontalAlignment = "Center"
   Name = "btnReset"VerticalAlignment = "Center"Width = "75"/>
</Grid>
</Border>
```

```
<!-- 忙碌指示控件 -->
<toolkit:BusyIndicator x:Name = "busy" IsBusy = "False" MinWidth = "0"/>
</Grid>
```

（5）引用服务，实现登录功能。选择 Silverlight 项目，右击，选择添加服务引用（出现一个窗体，单击出现当前工程中所有的 WCF 服务）。如图上机 6-4 所示。

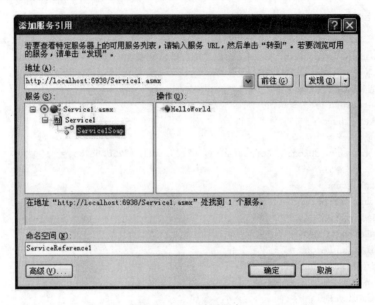

图上机 6-4　引用服务

在 Login.xaml.cs 类中调用服务的方法实现登录，代码如下：

```
//登录
Private void btnOK_Click(object sender, RoutedEventArgs e)
{
    String name = "this.txtName.Text.Trim()";
    String pwd = this.txtPwd.Password.Trim();
    If(name = "" || pwd = "")
    {
    MessageBox.Show("用户名和密码不能为空!");
    Return;
    }
    //调服务
    Try
    {
    LoginServiceClient client = new LoginServiceClient();
    clinet.LoginCompleted += new
EventHandler<LoginCompletedEventArgs>
    (OnLoginCompleted);
    client.LoginAsync(name,pwd);
    }
```

```
            Catch(Exception)
          {
            MessageBox.Visibilty = Visibilty.Collapsed;
             //显示忙碌指示控件
             this.busy.IsBusy = true;
          }
            Private void OnLoginCompleted(object sender,LoginCompletedEventArgs e)
          {
            This.busy,IsBusy = false;
            This.dbLogin.Visibility = Visibility.Visible;
            If(e.Error!= null)
            {
          MessageBox.Show(e.Error.Message);
          Return;
          }
            If(e.Result > 0)
          {
             //登录成功后的主页面
             MainPage mian = new MainPage();
       //图片
          Image imgBg = new Image();
          BitmapImage bit = new BimapImage();
          Bit.UriSource = new Uri("Imgs/3.jpg",Urikind.Relative);
          imqBg.Source = bit;
          imgBg.Opacity = 0.3;
          mian.EmpName = this.txtName.Text.Trim();
           mian.right.Child = imgBg;
           this.Content = mian;
       }
       Else
       {
          MessageBox.Show("用户名和密码错误或账户已离职");
       }

       }
```

指导 2 实现员工管理的新增

完成本任务所用到的主要知识点：
- WCF 服务
- Linq To Sql 使用

问题

前面创建数据库以及数据表，用于实现收银管理。现在要对员工进行管理，效果如图上机 6-5 所示。

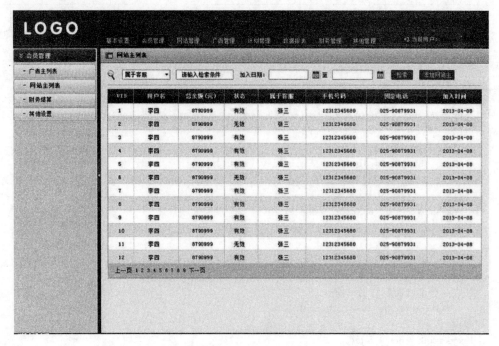

图上机 6-5　员工管理页面

分析

通过在没有选中数据表格中的数据时刻，我们可以在查看区对员工进行增加。

解决方案

(1) 添加员工的 WCF 服务（EmpService.svc）。

```
//< summary >
//添加员工
//</ summary >
//< param name = "no"></ param >
//< param name = "name"></ param >
//< param name = "pwd"></ param >
//< param name = "roNo"></ param >
//< param name = "state"></ param >
//< returns ></ returns >
[OperationContract]
Public int EmpAdd(string no, string name, string pwd, string roNo,
 String state)
{
 //0:已存在(员工号); -1:异常;1:添加成功
 Var emps = from e in entities.empInfo where e.eNo == no select e;
 If(emps.Count()> 0)
 {
  Return 0;
```

```
        }
    Else
      {
       Try
        {
          empInfo emp = new empInfo();
          emp.eNo = no;
          emp.eName = name;
          emp.Pwd = GetHashCode(pwd);
          //注意:是否能成功添加到数据库
         //报错:roNo 是对象键部分信息,不能修改
         //emp.roleInfo.roNo = roNo; //改为下行
          emp.eRoleNo = roNo;
          Emp.state = state;
          //添加对象到实体集
          entities.empInfo.AddObject(emp);
          //保存更改到数据库
          entities.SaveChanges();
           {
             Return 1;
           }
          Catch(Exception)
           {
             Return -1;
           }
         }
       }
```

(2)引用服务并实现员工的新增,代码如下:

```
XAML:
< Grid x:Name = "LayoutRoot" Background = "Transparent"
 Loaded = "LayoutRoot_Loaded">
< Grid.RowDefinitions >
 < RowDefinition Height = "40"/>
 < RowDefinition Height = "40"/>
 < RowDefinition Height = " * * "/>
</Grid.RowDefinitions >
< Grid.ColumnDefinitions >
 < ColumnDefinition Width = " * ">
 < ColumnDefinition Width = "200">
</Grid.ColumnDefinitions >
<!-- title -->
 < TextBook Grid.Row = "0"Grid.ColumnSpan = "2"Text = "员工信息管理">
 VerticalAlignment = "Center"HorizontalAlignment = "Center"
   FontSize = "24"/>
<!-- 操作 -->
< StackPanel Grid.Row = "1"Grid.ColumnSpan = "2"VerticalAlignment = "Center"
```

```xml
    Orientation="Horizontal" HorizontalAlignment="Center">
      <TextBlock VerticalAlignment="Center" Text="请选择查询条件:"/>
      <ComboBox x:Name="cboEmp" Width="60">
        <ComboBoxItem Content="编号" Tag="eNo"/>
        <ComboBoxItem Content="姓名" Tag="eName"/>
      </ComboBox>
      <TextBox x:Name="txtEmpSerch" Width="60" Mrgin="10000" MaxLength="10"/>
      <Button x:Name="btnSerach" Width="75" Content="查询">
        <Click="btnSearch_Click"/>
      <Button x:Name="btnSearchAll" Width="75" Content="查询所有">
        <Click="btnSearchAll_Click"/>
    </StackPanel>
    <!--数据显示区域-->
    <sdk:DataGrid SelectionChanged="dgEmp_SelectionChanged"
      AutoGenerateColumn="False" FrozenColumnCount="2" x:Name="dgEmp"
      Grid.Row="2" Grid.Column="0">
      <sdk:DataGrid.Columns>
    <!--<sdk:DataGridCheckBoxColumn Header="复选"/>-->
    <sdk:DataGridTextColumn Header="员工编号" MinWidth="65"
      Binding="{Binding eNo}"/>
    <sdk:DataGridTextColumn Header="员工姓名" Min Width="65"
    Binding="{Binding.eName}"/>
    <!--<sdk:DataGridTextColumn Header="密码" Min Width="60" Width="60"
      Binding="{Binding pwd}">-->
    <!--员工角色-->
    <sdk:DataGridTextColumn Header="角色" Min Width="60" Width="*"
    Binding="{Binding state}"/>
    </sdk:DataGrid.Columns>
    </sdk:DataGrid>
    <!--操作数据-->
    <ScrollViewer Grid.Row="2" Grid.Column="1">
      <Grid>
    <Grid.RowDefinitions>
      <RowDefinition Height="30"/>
      <RowDefinition Height="160"/>
      <RowDefinition Height="40"/>
    </Grid.RowDefinitions>
    <!--title-->
    <Border Grid.Row="0" Grid.ColumnSpn="3" BorderThickness="5"
      CornerRadius="10">
    <TextBolck Text="查看区" VerticalAlignment="Center"
      HorizontalAlignment="Center" FontSize="14/>
    </Border>
    <!--详细信息-->
    <border borderThickness="5" CornerRadius="10" Grid.Row="1"
    Grid.ColumnSpan="3">
    <Grid>
    <Grid.RowDefinitions>
      <RowDefinition Height="30"/>
```

```xml
        <RowDefinition Height="30"/>
        <RowDefinition Height="30"/>
        <RowDefinition Height="30"/>
        <RowDefinition Height="30"/>
    </Grid.RowDefinitions>
    <Grid.ColumnDefinitions>
        <ColumnDefinition Width="1*"/>
        <ColumnDefinition Width="2*"/>
    </Grid.ColumnDefinitions>
    <TextBlock Text="员工编号":VerticalAlignment="Center"/>
    <TextBox x:Name="txtEmpNo"Grid.Column="1"Height="24"
        MaxLength="20"Margin="0,3,0,0"VerticalAliginment="Top"/>
    <TextBolck x:Name="txtEmpName"Grid.Row="1"Grid.Column="1"
        Height="24"MaxLength="10"Margin="0,3,0,0"
    VerticalAlignment="Top">
    <TextBolck Text="密码:员工初始密码为123"Grid.Row="2"
        Grid.ColumnSpan="2"VerticalAligint="Center"/>
    <!--<TextBox x:Name="txtPwd"Grid.Row="2"Grid.Column="1"
        Height="24"MaxLength="6"/>-->
    <TextBlock Text="角色:"Grid.Row="2"Grid.Column="1"
        Height="24"MaxLength="10"Margin="0,3,0,0"
        VerticalAlignment="top"/>
    <TextBlock Text="密码:员工初始密码为123"Grid.Row="2"
        Grid.ColumnSpan="2"VerticalAlignment="Center"/>
    <!--<TextBox x:Name="txtPwd"Grid.Row="2"Grid.Column="1"
        Height="24"MaxLength="6"/>-->
    <TxetBlock Text="角色:Grid.Row="3"VerticalAlignment="Center"/>
    <!--下拉列表框绑定数据-->
    <ComboBox x:Name="cboRole"Grid.Row="3"Grid.Column="1"
        SelectValuePath="roNo"DisplayMemberPath="roName"/>
    <TextBlock Text="状态:"Grid.Row="4"VerticalAlignment="Center"/>
    <!--员工状态-->
    <!--<TextBox x:Name="txtState"Grid.Row="4"Grid.Column="1"
        Height="24"/>-->
    <ComboBox x:Name="cboState"Grid.Row="4"Grid.Column="1">
        <ComboBoxItem Content="在职"/>
        <ComboBoxItem Content="离职"/>
    </ComboBox>
</Grid>
</Border>
<!--增删改-->
<Border Grid.Row="2"Grid.ColumnSpan="3"BorderThickness="5"
    CornerRadius="10">
<stackPanel Orientation="Horizontal"HorizontalAlignment="Center">
    <Button Content="添加"Height="23"Name="btnAdd"Width="35"
        Click="btnAdd_Click"/>
    <Button Content="删除"Height="23"Name="btnDel"Width="35"
        Click="btnDel_Click"Margin="5050"/>
    <Button Content="修改"Height="23"Name="btnAlt"Width="35"
```

```
        Click = "btnAlt_Click"/>
      <Button Content = "重置密码"Height = "23"Name = "btnReset"Width = "35"
       Click = "btnReset_Click"Margin = "5050"/>
           </StackPanel>
        </Border>
   </Grid>
</ScrollViewer>
<! -- 忙碌控件 -->
<toolkit:BusyIndicator x:Name = "busy"IsBusy = "False"Margin = "05050"/>
</Grid>
 C#：
 //根元素加载事件
Private void LayoutRoot_Loaded(object sender,RouteEventArgs e)
{
 InitEmps();
 InitCombox();
}
//实例服务客服端对象
 EmpServiceClient client = null;
//初始化员工数据
Private void InitEps()
{
 Try
 {
Client = new EmpServiceClient();
client.GetEmpByTypeCompleted + =
 New EventHandler<GetEmpByTypeCompletedEventArgs>(OnGetCompleted);
 client.GetEmpByTypeAsync("","");
  }
  Catch(Exception)
  {
 MessageBox.Show("调用服务失败!");
 Return ;
}
 this.busy.IsBusy = true;
}
Private void OnGetCompleted(object sender,GetEmpByTypeCompletedEventArgs e)
{
 this.busy.IsBusy = false;
 If(e.Error!= null)
{
MessageBox.Show(e.Error.Message);
Return;
}
 this.dgEmp.ItemSource = e.Result;
 If(e.Result.Count == 0)
   {
MessageBox.Show("没有匹配的数据");
 Return;
```

```
}
}
角色服务
RoleServiceClient roleClient = null;
//下拉列表框绑定数据
Public void InitComBox()
{
 roleClient = new RoleServiceClient();
 roleClient.GetRoleInfoCompleted +=
 New EventHandler<GetRoleInfoCompletedEventArgs>(OnGetRoleCompleted);
 roleClient.GetRoleInfoArgs();
}
Private void OnGetRoleCompleted(obj sender,GetRoleInfoCompletedEventArgs e)
{
 If(e.Error!=null)
 {
 MessageBox.Show(e.Error.Message);
 Return;
 }
 this.cboRole.ItemSource = e.Result;
}
//查询
Private void btnSearch_Click(object sender,RoutedEventArgs e)
{
 if(txtEmpSearch.Text.Time() = ""||cboEmp.SelectedItem == null)
 {
    MessageBox.Show("请确认是否输入了查询条件!");
    Return;
 }
Else
{
ComboBoxItem item = this.cboEmp.SelectedItem as ComboBoxItem;
String type = "item.Tag.ToString()";
String content = this,txtEmpSerach.Text.Trim();
//调用服务
Client = new EmpServiceClient();
client.GetEmpByTypeCompleted +=
New EventHandler<GetEmpByTypeCompletedEventArgs>(OnGetCompleted);
client.GetEmpByTypeAsync(type,Content);
this.busy.IsBusy = true;
   }
}
//查询所有
Private void btnSearchAll_Click(object sender,RoutedEventArgs e)
{
 InitEmps();
}
//添加
Private void btnAdd_Click(object sender,RouteEventArgs e)
```

```csharp
{
  //获取数据
  String no = this.txtEmpNo.Text.Trim();
  String name = this.txtEmpName.Text.Trim();
  String pwd = "123";
//string state = this.txtState.Text.Trim();
String state = ((comboBoxItem)(this.cboState.SelectedItem)).Content.ToString();
String roNo = this.cboRole.SelectedValue.ToString();
//验证
If(string.IsNullOrEmpty(no)||string.IsNullOrEmpty(name))||
string.IsNullOrEmpty(pwd)||string.IsNullOrEmpty(state)||
 string.IsNullOrEmpty(roNo)
{
 MessageBox.Show("请录入完整信息!");
 Return;
}
//调用服务
  Try
   {
Client = new EmpServiceClient();
clinet.EmpAddCompleted +=
New EventHandler<EmpAddCompletedEventArgs>(OnAddCompleted);
client.EmpAddasync(no,name,pwd,roNo,state);
   }
 Catch(Exception)
  {
    MessageBox.Show("调用服务失败!");
  Return ;
}
this.busy.IsBusy = true;
Private void OnAddCompleted(object sender,EmpAddCompletedEventArg e)
{
this.busy.IsBusy = false;
//0:已存在(员工),-1:异常,1:添加成功
If(e.Error!= null)
{
 MessageBox.Show(e.Error.Message);
 Return;
}
Switch(e.Result)
{
 Case 0:
  {
     MessageBox.Show("员工号已存在!");
     Break;
  }
  Case -1:
  {
 MessageBox.Show("服务出现异常,请联系×××");
```

```
    Break;
}
Case 1:
{
//刷新数据
 InitEmps();
 MessageBox.Show("添加成功");
 Break;
}
Default:
Break;
 }
}
```

第 2 部分 练习

练习 实现员工管理的删除和修改

问题

结合以上所学的知识点，使用 WCF 和 Linq 完成对员工的修改和删除。

➢ 选择一条已经存在的数据，在右边查看区会显示当前选中信息。
➢ 单击修改，对员工信息进行更正。
➢ 验证更新的信息输入是否有误。
➢ 删除员工。